BIRTH, EVOLUTION AND DEATH OF STARS

James Lequeux
Paris Observatory, France

BIRTH, EVOLUTION AND DEATH OF STARS

Published by

World Scientific Publishing Co. Pte. Ltd.
5 Toh Tuck Link, Singapore 596224
USA office: 27 Warren Street, Suite 401-402, Hackensack, NJ 07601
UK office: 57 Shelton Street, Covent Garden, London WC2H 9HE

Library of Congress Cataloging-in-Publication Data
Lequeux, James.
[Naissance, évolution et mort des étoiles. English]
Birth, evolution and death of stars / James Lequeux.
p. cm.
ISBN 978-9814508773 (pbk. : alk. paper) -- ISBN 9814508772 (pbk. : alk. paper)
1. Stars--Evolution. I. Title.
QB806.L4713 2013
523.8'8--dc23
2013009944

British Library Cataloguing-in-Publication Data
A catalogue record for this book is available from the British Library.

Originally published in French as "**Naissance, évolution et mort des étoiles**" by EDP Sciences.
 A co-publication with EDP Sciences, 17, rue du Hoggar, Parc d'activités de Courtaboeuf BP 112, 91944 Les Ulis Cedex A, France.

This edition is distributed worldwide by World Scientific Publishing Co. Pte. Ltd., except France.

Typeset by Stallion Press
Email: enquiries@stallionpress.com

Printed in Singapore by Mainland Press Pte Ltd.

Contents

Introduction

Astrophysics is a recent science. François Arago was the first to demonstrate, in 1811, that the surface of the Sun is not solid or liquid, but made of an incandescent gas. Later, he showed that this is also the case for stars in general. In 1860, Robert Bunsen and Gustav Kirchhoff discovered lines of several terrestrial elements in the spectrum of the Sun, marking what is often considered as the beginning of astrophysics. Then the progress of this science followed the progress of physics, until Hans Bethe and Carl Friedrich von Weiszäcker established the origin of stellar energy, immediately before World War 2. As to the evolution of stars, it only began to be understood in the 1950's. There are still some mysteries in their birth and death.

The present book describes, in a way that we hope is easily accessible to someone with a basic knowledge of physics, the various aspects of the formation, nature, evolution and death of stars. Although it is not a textbook for specialists, it contains a few formulae and simple demonstrations, when they are considered more telling than a popular account. I consider them as necessary for a good understanding of stellar physics.

The first chapter describes in a succinct way the interstellar medium from which stars are formed, and the different stages of their formation until the nuclear reactions which provide energy to the stars are stabilised. Nothing was known about these matters before radiotelescopes and space observatories allowed observations at radio and infrared wavelengths. Recent such observations are still providing us with many surprises.

Then the principles of the physics of stars are described. Here, there are less mysteries than in the previous chapter: indeed, it is possible to reproduce satisfactorily by numerical models the observed properties of stars, but one has to fix more or less arbitrarily some parameters which describe empirically poorly known physical processes like turbulence or convection. As a consequence, it is necessary to compare continuously the results of models to observations, as we will do.

The third chapter describes the evolution of stars with time, starting with that of the Sun. This evolution depends much of the mass of the star. High-mass stars burn their fuel frantically and live only a few million years, while low-mass ones, more sparing, can live several billion years.

However, all stars die when their usable fuel is exhausted. Their death is treated in Chapter 4. It is relatively quiet for low-mass stars, but high-mass stars die in a spectacular way through a gigantic explosion, whose physics is not yet fully understood.

Stars are not always isolated: about half are members of a double or multiple systems. We will see what happens to them in Chapter 5. It is only after space vehicles initiated ultraviolet, X-ray and gamma-ray astronomy that the strange and spectacular phenomena produced by mass transfer between the components could be observed. Everything is not yet understood in this domain, where discoveries succeed each other rapidly.

Like stars, galaxies evolve, and their evolution depends on their stellar content. Indeed, stars consume continuously interstellar gas, which they eject in part at the end of their lives, enriched in heavy elements produced by nucleosynthesis in their depths or during their final explosions. Then other stars are born from this gas which they further enrich, while the matter they have not been able to eject subsists forever as inert remnants. This evolution is described in the last chapter, where we will see how it can be observed.

We hope that the reader will have as much pleasure in discovering the fascinating world of stars as we had in writing this book.

Acknowledgements

Jean-Paul Zahn has been kind enough to read the entire manuscript. I am very grateful to him for this. I also wish to thank André Maeder, Klaas de Boer and Wilhelm Seggewiss for allowing me to use figures from their books.

Insert : Stellar nomenclature.

From antiquity to the 17th century, stars were only designated by names. This usage is still preserved for the brightest ones, with names generally of Arabic origin like Sirius, Aldebaran, Betelgeuse, etc. Some more recent names are also used, for example Mira Ceti which is a particularly remarkable variable star in constellation of the Whale (Cetus). However, this nomenclature was soon considered as insufficient and unpractical. In 1601, the German astronomer Johann Bayer, in his celestial atlas *Uranometria*, designated stars by a Greek letter followed by the genitive of Latin name of the constellation to which they belong. In principle, α is the brightest star of the constellation (for example α Cygni = Deneb), followed by β, γ, etc. But there are in general more stars visible with the naked eye in a constellation than letters in the Greek alphabet, so that in the 18th century the English astronomer John Flamsteed introduced another system, in which the stars of a given constellation are designated by a number increasing with right ascension. This system is also in use: for example, 51 Pegasi is the first star around which an exoplanet has been discovered. But the astronomers who constructed large stellar catalogues from the beginning of the 19th century, for example Lalande who catalogued 50 000 stars in 1801, found this nomenclature still insufficient and difficult to use. Instead, they generally classed the stars in order of right ascension without taking the constellations into account. This is the case for the famous Henry Draper (HD) catalogue. The Bonner Durchmusterung (BD) catalogue, which contains a very large number of stars, is divided into 1° zones in declination; star are listed by right ascension inside each zone. A further complication concerns variable stars, which are often designated by one or two Latin capital letters preceding the name of the constellation (example: T Tauri). Moreover, the number and the limits of the constellations have changed with time: it was only in 1930 that the International Astronomical Union fixed in a definitive way the limits of the constellations, introducing some changes in doing so. At present, one tends to designate stars and other celestial objects by their position in the sky (right ascension and declination); it is necessary in this case to specify the date for these coordinates, which change with time due to precession. When the letter B, or no letter, precedes the coordinates, this means that this date is 1950,0, while the letter J indicates a date of 2000,0.

A great amount of confusion results from these different systems. A given star can possess a large number of different names: for example, Betelgeuse is designated by about 40 names, of which the following are examples: α Orioni, 58 Orioni, HD 39801, BD+07 1055, IRAS 05524+0723, 2MASS J055108+0724255, etc.

In order to find one's way through this mess, the best is to access the free database SIMBAD of the *Centre de Données de Strasbourg* (http://simbad.u-strasbg.fr/) where one can find from any one of the names all the different designations of a star and the catalogues from which they originate. As a bonus a large number of data about the target star is also available.

1

The Birth of Stars

1.1 Interstellar matter

1.1.1 The different phases of interstellar matter

Stars are born from interstellar matter. This matter consists of gas and dust. The dust grains are made of silicates or carbonaceous matter, and can be covered with ices of water, carbon monoxide or other frozen products; they make up 1 to 2 per cent of the mass of the interstellar matter. They contain essentially elements heavier than hydrogen or helium: as a consequence, the interstellar gas is deficient in heavy elements with respect to the chemical composition of the Galaxy near the Sun, which is given in Table 1.1. This deficiency has very important consequences for the physics of interstellar matter, but has no consequences for the chemical composition of stars: as a matter of fact, when stars form from the interstellar matter, they engulf dust as well as gas.

There exists several phases in the interstellar medium, as determined by density and by the nature and intensity of the ambient radiation.

Except in the densest regions, the interstellar medium is subject to stellar radiation. Those stellar photons whose wavelength is shorter than 91.2 nanometers (nm) ionize hydrogen, which is by far the most abundant constituent of interstellar matter. As the ionisation cross-section of hydrogen is very large, this ionisation is in practice a yes or no phenomenon.

Element (X)	$12 + \log(n_X/n_H)$	Element (X)	$12 + \log(n_X/n_H)$
H	12.00	**Si**	7.27
He	10.99	**P**	5.57
Li	3.31	**S**	7.09
C	8.56	**Ar**	6.56
N	7.97	**K**	5.13
O	8.77	**Ca**	6.34
Ne	8.03	**Ti**	4.93
Na	6.31	**Fe**	7.50
Mg	7.40	**Ni**	6.25

Table 1.1. Chemical composition of the Galaxy near the Sun; only the most interesting elements are listed. n_X and n_H are respectively the number of atoms of element X and of hydrogen, which is the reference. As usual, we give the abundances with respect to hydrogen in a logarithmic scale increased by 12. The proportions in mass are approximately: hydrogen: $X = 0.70$; helium: $Y = 0.28$; heavy elements: $Z = 0.02$.

Therefore, hydrogen is almost entirely ionised in a volume around a hot star. Outside this volume, the radiation coming from the star is deprived of photons with wavelength shorter than 91.2 nm by this ionisation. As a consequence, there are in the interstellar medium regions where hydrogen is 100% ionised: these are the *gaseous nebulae*, also designated as *HII regions*[1]. While their size increases with time due to the internal pressure of their warm gas (at about 10^4 K), these nebulae eventually fill the surrounding space with a tenuous gas, which survives for a long time because ions and electrons are slow to recombine: this is the *diffuse ionised medium*. The regions where hydrogen is essentially neutral and atomic are the *HI regions*, for which a large range of densities and temperature exists, from relatively cold and dense clouds to a warm, diffuse medium. If the density of the neutral medium is large, say more than 1 000 atoms per cm^3, hydrogen is molecular rather than atomic: we speak in this case of *molecular clouds* (Figure 1.1). Finally, there exists regions filled with very hot (about 10^6 K) and very low-density (some 10^{-3} ion per cm^3) ionised gas;

[1] Astronomers usually designate as follows the ionization degrees of an element X. Neutral: **X**I, once ionised: **X**II, twice ionised: **X**III, etc. Neutral hydrogen H^0 is therefore HI, and ionised hydrogen, H^+, HII.

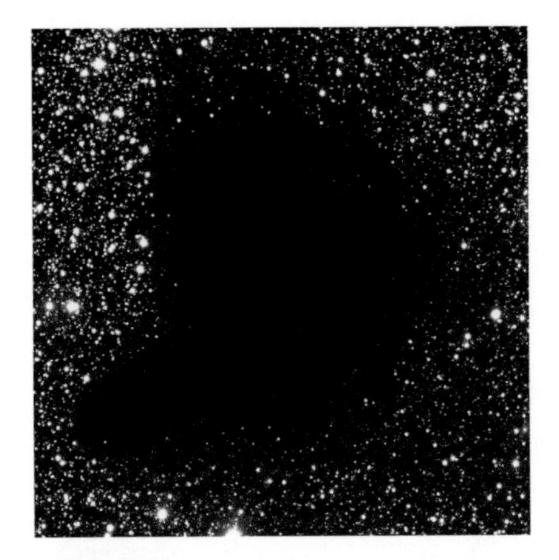

Figure 1.1. The molecular cloud B68. Left, an image in visible light taken with one of the 8-m diameter telescopes of the Very Large Telescope (VLT) of the European Southern Observatory (ESO). The visible light of the background stars is totally obscured by dust contained in the cloud. Right, a composite image in false colours obtained in the infrared at wavelengths 1.25 μm (blue), 1.65 μm (green) and 2.17 μm (red). Infrared extinction being smaller than the visible one, we can now see stars inside or behind the cloud, especially at 2.17 μm. © *ESO.*

this gas has been ejected by stars at the end of their lives. These *very hot regions* occupy a larger volume than all the other phases together, but do not contain much mass.

The different phases of the neutral interstellar matter are approximately in pressure equilibrium with each other and with the very hot gas. This pressure is proportional to the product of the density n and the absolute temperature T, with nT of the order of 3 000 atom cm^{-3} K in the solar vicinity. Conversely, pressure is much higher in HII regions, because their temperature is of the order of 10 000 K, so that they are in expansion and are surrounded by an ionisation front as well as often by a shock. Table 1.2 summarises the properties of the interstellar matter in our Galaxy.

In neutral regions, there are ultraviolet photons with wavelength larger than 91.2 nm which are able to pull out electrons from dust grains or from those atoms whose ionisation energy is smaller than that of hydrogen (13.6 electron-volts, eV): carbon, sulphur and all metals, but not helium, nitrogen and oxygen. This has great importance for the physics of the medium. The heating of HI regions results from the ionisation of dust: the extracted electrons, which have an energy of the order of 10 eV, come into

Medium		Density (atom cm^{-3})	Temperature (K)	Total mass (solar masses)
Atomic (HI)	Cold	≈ 25	≈ 100	1.5×10^9
	Warm	≈ 0.25	$\approx 8\,000$	1.5×10^9
Molecular		$\geq 1\,000$	$5 - 100$	2×10^9?
Ionised	HII regions	$1 - 10^4$	$\approx 10\,000$	5×10^7
	Diffuse	≈ 0.03	$\approx 8\,000$	10^9
	Hot	$\approx 6 \times 10^{-3}$	$\approx 5 \times 10^5$	10^8?

Table 1.2. Interstellar matter in our Galaxy.

thermal equilibrium with the electrons already present, which they heat. Then electrons and the other particles come into thermal equilibrium more slowly. In the regions of particularly low density of the neutral medium, another source of heating is the X-ray radiation emitted by supernova remnants and close binary stars. The cooling of the neutral medium is mainly due to the emission of the line of ionised carbon CII at the wavelength of 158 μm.

The interstellar medium is constantly agitated by the mechanical effect of the explosions of the most massive stars at their death (supernovae) and of the violent winds emitted by high-mass stars, which are very much more intense than the solar wind. This generates turbulence and a fractal structure which can be seen in radio maps of the HI medium obtained in the 21-cm line of neutral hydrogen, or in the maps of the molecular gas obtained in the lines of the CO and other molecules (Figure 1.2), or finally in the thermal emission of dust in the far infrared. The galactic magnetic field also affects the medium, which is always more or less a conductor of electricity and is therefore sensitive to this field. Gravity also plays a role, producing instabilities and condensations in the interstellar matter, which form stars by gravitational collapse. We will therefore examine more closely the physics of these dense regions, because they are the places where stars are formed.

1.1.2 Molecular clouds

The heating and cooling processes in molecular regions are completely different from those in the HI medium. One reason is that ultraviolet photons

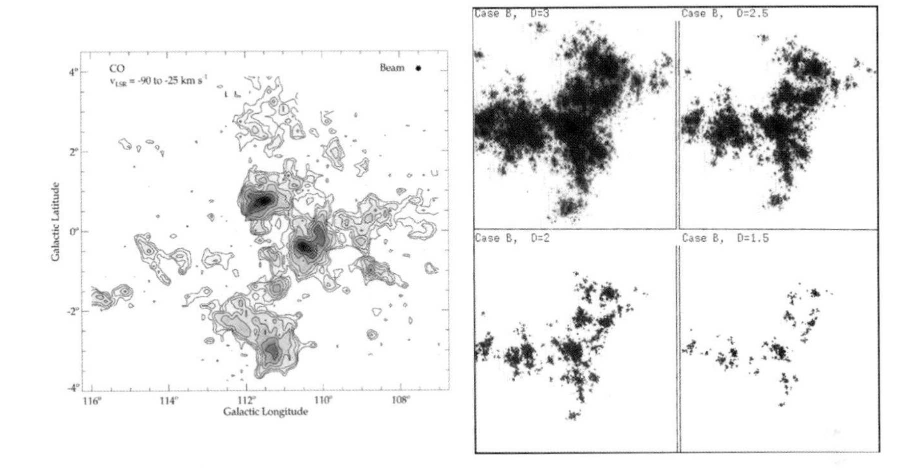

Figure 1.2. Left, the map of a complex of molecular clouds obtained in the line of the CO molecule at 115 GHz (wavelength 2.6 mm). Right, simulation of a fractal cloud with the fractal 3-D dimension D.[1] The simulation and the observed cloud are globally similar for $D = 2$ to 2.5. The same considerations apply whatever the considered scale, a fact which characterises a fractal structure. *From Ungerecht, H. et al. (2000) Astrophysical Journal 537, 221, with permission of AAS, and Pfenniger, D. & Combes, F. (1994) Astronomy & Astrophysics 285, 94, with permission of ESO.*

cannot penetrate these regions, because they are absorbed by the dust at the edge of the cloud; another one is that the matter is under a different form. There are no more isolated atoms, except for the noble gases, because they are combined into molecules which are present either as ices at the surface of dust grains or in gaseous form: one has observed more than 150 different types of molecules through their millimetre and submillimeter rotation lines, not counting symmetrical molecules which are unobservable in this way because their rotation transitions are forbidden by selection rules. The heating is now due to cosmic rays, i.e. charged particles with energies of a few million electron-volts (MeV) which can propagate inside these dense clouds. Their collisions with molecules pull out electrons which come into thermal equilibrium with the electrons already present and heat the gas in this way. In the deepest parts of the molecular clouds where even cosmic rays cannot penetrate, the only possibility to heat the gas is the collisions

[2] The fractal dimension D is such that the number $N(>L)$ of structures with a size larger than L is given by the expression $N(>L) \propto L^{-D}$.

of molecules with the dust grains, which themselves are very difficult to heat for lack of radiation: the temperature is extremely low there, only a few K. The cooling mechanism is the emission of rotation lines of molecules, especially those of CO which is particularly abundant. The dipole transitions of molecular hydrogen are forbidden because of the symmetry of this molecule, and its quadrupole lines are extremely weak and do not participate in the cooling of molecular clouds, although molecular hydrogen is by far the most abundant molecule in these clouds.

Of course, there are transition regions between molecular clouds and the external medium. Hot stars which are strong emitters of ultraviolet radiation are often present near these clouds. Their radiation penetrate more or less deeply inside the cloud according to their energy, which results in a structure with layers of different composition, as represented on Figure 1.3. These *dissociation regions*, or *photon dominated*

Figure 1.3. The photodissociation region near the HII region M 17. This image in false colour was obtained in the near infrared at 1.25 μm (blue), 1.65 μm (green) and 2.17 μm (red), with the New Technology Telescope (NTT) of ESO. To the right, there is a totally opaque molecular cloud: the few visible stars are in front. To the left, the brightest stars are hot stars which ionise the gas of M 17. In between is the very bright photodissociation region, whose emission is dominated by the continuum of the ionised gas it contains. Its structure is very fragmented. © *ESO.*

regions (PDR), receive a lot of energy from the stars. This energy which is to a large extent re-radiated in several spectral lines, in particular the line of CII at 158 μm in the far infrared, that of neutral oxygen OI at 63 μm and that of neutral carbon CI at 370 μm. Rotation-vibration and even pure rotation lines of molecular hydrogen in the near and mid-infrared can also be observed, as well as many molecular lines at millimetre and submillimeter wavelengths. Finally, the dust grains, which are heated by the ultraviolet radiation, emit strongly in the infrared. The PDRs thus play an important role in the energy balance of our Galaxy. They are also places with a very active chemistry, which differs from that in molecular clouds because the temperature can reach higher values, of the order of 1 000 K.

The preceding discussion is actually somewhat simplistic, given the extreme complexity and fragmentation of the interstellar medium. The main effect of this fragmentation for the physics of PDRs is to allow ultraviolet radiation to penetrate deeper inside molecular clouds, increasing as a consequence the thickness of the surrounding photodissociation region. But it does not affect much what occurs in their interiors.

1.2 Star formation

Stars form by collapse of the densest parts of molecular clouds. Observations indicate that the coldest clouds tend to form low-mass stars, observed first in the infrared inside the clouds, then in visible light at their surface when the clouds dissipate (Figure 1.4). Giant molecular clouds, which are generally warmer, produce stars of all masses. The most massive can be seen in the infrared before they have dissipated their dense parent cloud; they can also be observed through secondary phenomena, the most remarkable of which are the natural masers produced by some molecules, OH at a wavelength of 18 cm and H_2O at 1.35 cm. The formation of massive stars appears to be a contagious phenomenon (Figure 1.5). The pressure and shock waves due to stellar winds and final explosions of previous-generation massive stars clearly trigger the formation of other stars in what remains of the cloud, or in

Figure 1.4. The Orion nebula and its cluster of young stars, seen in the near infrared. This image in false colour was obtained in the near infrared at 1.25 μm (blue), 1.65 μm (green) and 2.17 μm (red), with one of the 8-m diameter telescopes of the VLT of ESO. It reveals a rich cluster of nearly 1 000 young stars, with an age of about 1 million years. They are located under the surface of a molecular cloud which is located behind the HII region and is not visible optically. It is this HII region which produces the diffuse radiation. The ultraviolet radiation of the four bright stars at the centre which form the Orion Trapezium ionise the HII region and, progressively, the molecular cloud. © *ESO.*

nearby clouds. Star formation seems invariably accompanied by jets, often spectacular ones. Finally, recent observations from the HERSCHEL satellite show that stars tend to form in filaments (Figure 1.6), probably due to magnetic fields.

The theory of star formation is a difficult domain because of the complexity of molecular clouds, and also because the physical phenomena which take place are numerous and often poorly understood, in particular turbulence which very likely exists in those clouds. The role of the magnetic field is certainly important, but is unclear and controversial. Observations of molecular clouds are of little help, at least for the time being. All of interest that can be seen are the slow collapse motions of the

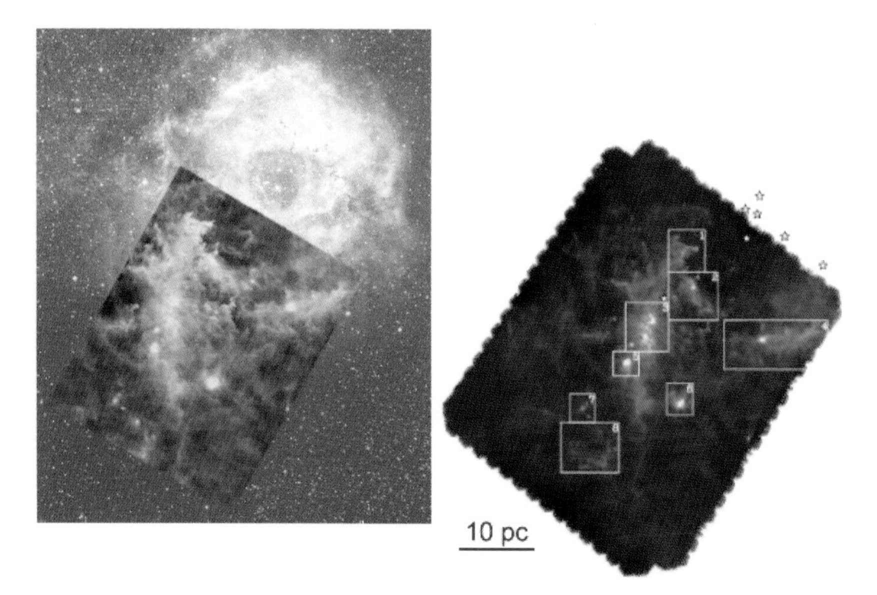

Figure 1.5. Contagious formation of stars near the Rosette nebula NGC 2244, observed with the HERSCHEL satellite. Left, an image of dust thermal emission in the far infrared superimposed on a photograph of the nebula. The false colours correspond to: blue, 70 μm; green, 160 μm; red, 500 μm. Dust being well mixed with the gas, what is seen is also the distribution of this gas. Right, an image at 250 μm, with the same scale. The star symbols indicate the position of the 7 stars which ionise the nebula. The rectangles delineate the regions of star formation where protostars are visible as bright points. Star formation is triggered by compression of the molecular cloud by the HII region. *From Schneider, N. et al. (2010) Astronomy & Astrophysics 518, L83, with permission of ESO.*

inner regions of these clouds which produce a broadening or splitting of some radio molecular lines through the Doppler–Fizeau effect.

It might be that stars in formation at places where accretion of matter is favoured grow at the expense of nearby protostars. As long as the privileged star grows, its advantage becomes larger as its gravitational field increases, and it can capture more and more material and even entire protostars. This mechanism could explain why giant molecular clouds form simultaneously a small number of massive stars, the privileged ones, and many low-mass stars which have not been captured, as can be seen in the Orion nebula (see Figure 1.4). Another possible scenario, which may coexist with the previous one, is that massive stars are formed in compact stellar groups by collisions and coalescence of less massive ones. These

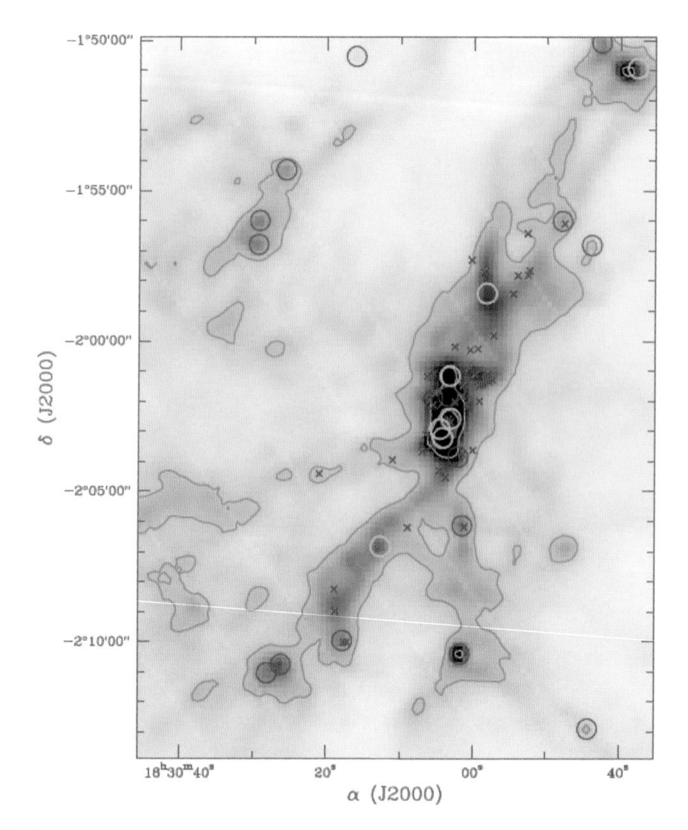

Figure 1.6. Formation of stars in constellation Serpens, observed with the HERSCHEL satellite. The dimensions of this image correspond to 1.3 × 1.8 pc at the distance of the region. The grey background and the contours indicate the distribution of the emission of interstellar dust at 350 μm. The molecular gas is distributed in the same way. The blue circles represent the location of protostar candidates (the warmest regions of the cloud), the green circles that of Class 0 protostars (still warmer) and the red crosses that of more evolved Class I protostars; the latter were observed in the mean infrared with the SPITZER satellite. One notices that star formation takes place in filaments, whose structure is probably determined by the interstellar magnetic field. *From Bontemps, S. et al. (2010) Astronomy & Astrophysics 518, L85, with permission of ESO.*

phenomena are still poorly understood and very difficult to check by observation, so we will not describe them any further. One may hope, however, that adaptive optics and interferometry will provide theorists with the observational data which are presently missing. In particular, one should be able to explain why the distribution of masses of stars at their birth (what we call the *Initial Mass Function*, IMF), seems to always be the same if one considers globally any extended region of our Galaxy or of external galaxies.

1.2.1 The gravitational collapse of molecular clouds

Before entering the subject, we will say a few words about the equilibrium and the structure of molecular clouds, supposing for the moment that they are not fragmented. The condition for equilibrium was first given by Clausius in 1870 and specified by Poincaré in 1911: it is the *virial equation*, which takes the following form if there is no magnetic field:

$$2E_{\mathrm{kin}} + E_{\mathrm{pot}} = 0, \tag{1.2}$$

E_{kin} is the kinetic energy of the system and E_{pot} its potential energy. For example, for a spherical cloud made of perfect gas, with a uniform temperature and without macroscopic motion, one has:

$$E_{\mathrm{kin}} = E_{\mathrm{therm}} = 3/2\ M\mathrm{k}T/\mu\mathrm{m_H}, \tag{1.3}$$

where M is the mass of the cloud, k the Boltzmann constant, T the temperature, μ the mean molecular mass (about 2.4 for a molecular cloud, taking into account heavy elements) and $\mathrm{m_H}$ the mass of the hydrogen atom. If there is macroscopic motion, one may in a first approximation add them quadratically to the thermal motions in order to obtain an equivalent temperature; however, this is dangerous if these motions are due to turbulence, especially supersonic turbulence as it is often the case. A magnetic field adds surface and volume terms, but we will ignore this complication here.

For a uniform density cloud, the potential energy is:

$$E_{\mathrm{pot}} = -3/5\ \mathrm{G}M^2/R, \tag{1.4}$$

G being the gravitation constant and R the cloud radius. The numerical factor is different if the density is not uniform in the cloud.

Gravitational instability was first studied by Jeans in 1902. We give here a demonstration due to Boer and Seggewiss (2008), which is better than the original one. A volume element is subject to the gravitational attraction of the inner regions of the cloud, i.e. $F_g = G\rho M/R^2$ where ρ is the specific mass per unit volume, and to the opposed force due to the gradient of the pressure P, $F_P = dP/dr$. For a perfect gas, $P = \rho T\mathfrak{R}/\mu$, $\mathfrak{R}$ being the perfect gas constant, hence $F_P = \mathfrak{R}T/\mu \, d\rho/dr$. The equation of motion is:

$$d^2r/dt^2 = -GM/R^2 - \mathfrak{R}T/\mu\rho \, d\rho/dr, \tag{1.5}$$

where the – sign means a motion towards the centre. There will be an instability if acceleration is negative, i.e. if $GM/R^2 > \mathfrak{R}T/\mu\rho/d\rho/dr$. Let us make the approximations $d\rho/dr \approx \rho/r$, and $R = (3M/4\pi\rho)^{1/3}$. The cloud will be unstable if its mass is larger than:

$$M_{Jeans} \approx (3/4\pi)^{1/2} (\mathfrak{R}T/\mu G)^{3/2} \rho^{-1/2}. \tag{1.6}$$

A more rigorous analysis gives the following numerical expressions:

$$M_{Jeans} = 100\ T^{3/2} n^{-1/2}\ M_\odot \text{ for the HI medium, and} \tag{1.7}$$

$$M_{Jeans} = 25\ T^{3/2} n^{-1/2}\ M_\odot \text{ for the molecular medium,} \tag{1.8}$$

where T is given in K and n is the number of atoms (or of molecules for the seond expression) per cm^3; 1 $M_\odot$ (solar mass) = 2×10^{33} g.[2]

However, molecular clouds are generally subject to an external pressure P_{ext}. In order to study stability, this pressure P_{ext} should be compared to the pressure under the surface of the cloud P_0. The latter always is for a uniform cloud:

$$P_0 = 3MkT/4\pi R^3\ \mu m_H - 3/5\ GM^2/4\pi R^4. \tag{1.9}$$

Differentiating this expression with respect to R, we can see that P_0 goes through a maximum when the value of the radius R_m is equal to:

$$R_m = 4/15\ GM\mu m_H/kT. \tag{1.10}$$

Let us assume that P_0 and P_{ext} are the same, i.e. that there is equilibrium. If $R > R_m$, a small compression will raise P_0 and the radius will adjust itself so that the pressures re-equilibrate: the equilibrium is stable.

---

[3] Like all astrophysicists, we will use cgs units; we apologise for this.

If, conversely, the radius is smaller than R_m, a small compression will decrease P_0, so that the cloud will contract further under the effect of the external pressure: the equilibrium is instable and the cloud collapses.

The instability of a cloud with radius $R < R_m$ with respect to an increase of external pressure explains why stars can form by contagion: the clouds located near a hot star can be affected by the increasing pressure of the ionisation front of the HII regions which surrounds the star, or, if the star exploded as a supernova, by the pressure of the shock wave resulting from this explosion.

One can consider Equation (1.9) in another way: if T and P_0 are fixed, it gives a relation between mass and radius, from which one deduces that for a given radius the mass of a stable cloud cannot be larger than some value M_{max}. For the realistic case of an isothermal cloud, which is not uniform as we will see later, one has:

$$M_{max} = 13[(P_{ext}/\mathrm{k})/3\,000\ \mathrm{K\ cm^{-3}}]\ (R/1\ \mathrm{pc})^2\ \mathrm{M}_\odot, \qquad (1.11)$$

with 1 pc = 3.08×10^{18} cm. If the mass is larger than this value, the cloud must fragment. It seems that interstellar clouds are generally close to condition (1.11). For example, there are in the Galaxy fragmented clouds whose mass and size is of the order of 10^6 $\mathrm{M}_\odot$ and 300 pc respectively. These clouds collapse when a crossing density wave increases the external pressure. There are also small clouds with radii of a few parsecs and masses of a few hundred $\mathrm{M}_\odot$ which can directly form a small number of stars.

Interstellar clouds are not uniform. An isothermal self-gravitating cloud at equilibrium cannot have a uniform density because the gravitational force depends on the radius r inside the cloud. This is also true in general for non-isothermal clouds. Density profiles of realistic clouds at equilibrium, subject to the standard interstellar radiation field and to an external pressure such that $nT = 3\,000$ K cm^{-3}, are shown on Figure 1.7. Their density varies approximately as $r^{-1.3}$, except in the central regions where it is uniform. The numerical factor of Equation (1.4), obtained for a uniform cloud, is thus incorrect, but this equation gives orders of magnitude sufficient for a discussion. Moreover, these models do not take into account the magnetic field and turbulence, which can change the results appreciably. In particular, turbulence can generate protostellar

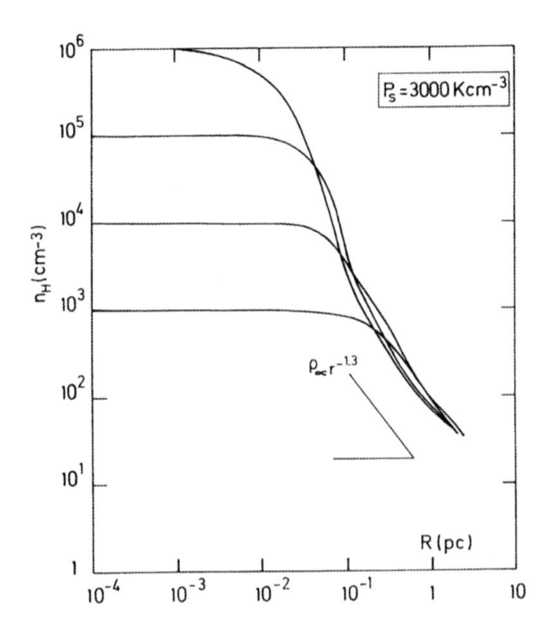

Figure 1.7. Density profiles for molecular clouds at equilibrium subject to an external pressure such that $nT = 3\,000$ K cm^{-3} and to a standard ultraviolet interstellar field. *From Chièze, J.-P. & Pineau des Forêts, G. (1987) Astronomy & Astrophysics 183, 98, with permission of ESO.*

condensations inside an apparently stable cloud, a property which is not fundamentally modified by a magnetic field. We cannot discuss these complications here.

Once instability sets in, for example due to an increase of external pressure, the cloud collapses upon itself. One can show that the characteristic time t_{collapse} for this collapse is roughly:

$$t_{\mathrm{collapse}} = (3G\rho)^{-1/2}, \tag{1.12}$$

ρ being the density, provided that all the gravitational energy is evacuated by radiation. This is a fast phenomenon at astronomical scales: for a cloud with a density of 5 000 H_2 molecules per cm^3, t_{collapse} is of the order of 4×10^5 years. In the unrealistic case of a cloud of uniform density at the beginning, all layers would arrive together at the centre. Figure 1.8 gives an example of collapse in a more realistic case: a nucleus with approximately uniform density forms, upon which matter falls from less deep layers, producing a shock at the surface of this core.

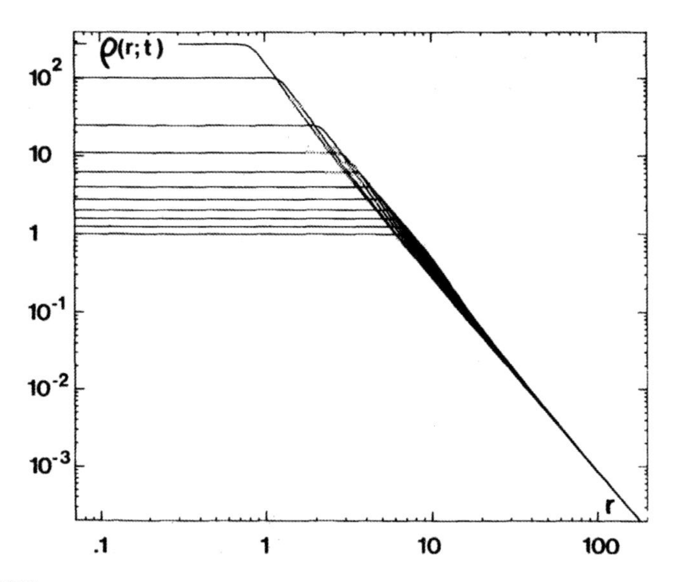

Figure 1.8. An example of an analytic solution of the collapse of a spherical cloud without magnetic field. The specific mass is plotted as a function of radius in a logarithmic scale for different values of the time. *From Blottiau P., Bouquet, S. & Chièze, J.-P. (1988) Astronomy & Astrophysics 207, 24, with permission of ESO.*

In these studies, fragmentation has been neglected. It can however occur during the collapse, for example due to turbulence, but one does not know for sure under which conditions. However it is certain that this fragmentation exists and is efficient, since stars tend to form in groups.

At the beginning of the collapse, the heat resulting from the contraction of the gas is evacuated through emission of molecular lines, in particular those of the CO molecule and of its isotopic substitutions ^{13}CO, $C^{18}O$ and $C^{17}O$, as well as through the thermal emission of dust in the far infrared. But there is a time when the optical thickness of the matter becomes sufficiently large, even in the far infrared, to stop the cooling by radiation. The medium then heats up continuously while its density grows: the contraction becomes adiabatic in the central regions.

Meanwhile, the dense core continues to grow due to the infall of the surrounding matter. If the mass is large enough, more than 0.07 times that of the Sun, nuclear reactions occur and a star is born. If the mass is smaller, an aborted star forms, a *brown dwarf*. We will say more about this later.

It appears however that the opaque fragments must have a minimum mass to yield stars. In effect, it is essential that the gravitational energy liberated in the collapse be radiated in order to allow the cloud to fragment, or in other words the contraction is not yet adiabatic. Let us write this condition for a homogeneous cloud. The rate of loss of gravitational energy of the collapsing cloud is of the order of $GM^2/Rt_{\text{collapse}}$, whilst the radiated energy is close to that of a blackbody with radius R and temperature T, i.e. $4\pi R^2\sigma T^4$, where σ is the Stefan–Boltzmann constant. The contraction will be adiabatic, so that the cloud will not fragment, if:

$$GM^2/Rt_{\text{collapse}} > 4\pi R^2\sigma T^4. \tag{1.13}$$

Taking t_{collapse} from Equation (1.12) and using Equation (1.5) without external pressure in order to obtain a relation between M, R and T, a rather simplistic approximation for a collapsing cloud, one obtains a minimum mass of the order of 0.07 $M_\odot$. This calculation is very rough; more sophisticated simulations show that the mass of the fragments cannot be smaller than 0.007 $M_\odot$, which is the characteristic mass of a very large planet. As a consequence, although brown dwarfs can be formed like ordinary stars from fragments of molecular clouds, this cannot be the case for planets which must be produced in another way, by coalescence of solid fragments and subsequent accretion of gas.

1.2.2 The angular momentum problem: circumstellar disks and jets

The preceding scenario only makes sense if the initial cloud has no rotation. If it rotates, even slowly, its angular momentum is preserved, so that the inner parts rotate faster and faster during contraction, which tends to oppose to the collapse; at least, this collapse can only take place along the rotation axis, so that the object becomes a rapidly rotating disk. One could imagine that planetary systems form in this way, with a central star surrounded by a rotating disk where planets are born. This is too simplistic: in the Solar System, most of the angular momentum is in the planets, while the Sun rotates slowly upon itself. This situation also occurs in the many planetary systems which have been observed around nearby stars. Some angular momentum has necessarily been transported from inside to outside, probably through the viscosity of the gas which is due to turbulence, magnetic field, or both. However the details are controversial. Let us assume, for an order of magnitude calculation, that the viscosity is large enough so that the cloud rotates like a solid body with an angular velocity ω. In the equatorial plane, the centrifugal force on a mass m at radius R is $F_c = m\omega^2 R$, and the gravitational force is $F_G = GmM/R^2$, M being the mass inside R. There is a critical radius R_{crit} where $F_c = F_G$:

$$R_{crit} = (GM/\omega^2)^{1/3}. \qquad (1.14)$$

The collapse of the inner regions with $R < R_{crit}$ cannot be stopped by rotation, while the external regions can only contract in a disk.

Numerical models confirm precisely this scenario. Moreover, they show that the region inside the critical radius accretes matter after its separation. For example, one simulation which concerns the collapse of a rotating cloud of mass 1 M⊙ foresees that after 45 000 years the mass of the star formed at the centre is 0.3 M⊙, and that the surrounding disk has a radius of about 100 a.u. (astronomical unit, the semi major axis of the terrestrial orbit, 150×10^6 km). When the star begins its adult life, after 190 000 years, its mass has grown to 0.6 M⊙ while the protoplanetary disk of mass 0.4 M⊙ has reached a radius of 1 000 a.u.: the central object has succeeded in capturing a part of the material of the disk. These values are in rough agreement with observations. However this model is limited by the assumption of axial symmetry. If we were to abandon this hypothesis, which is still useful in order to avoid gigantic calculations, one would probably see spiral structures in the disk, and perhaps a double or multiple star.

Indeed, another way to solve the angular momentum problem is to form a double or multiple star. The angular momentum is now in the revolution of the components around each other. This is a result obtained in some numerical simulations, less sophisticated than the previous one but with no axial symmetry. They suggest that the rotating disk which results from the collapse can fragment into two pieces with little rotation, each of which will become a star, or into even more pieces giving a multiple system (Figure 1.9). Actually, a very large fraction of stars (almost all the massive ones) belong to double or multiple systems. But even here the details are not well understood.

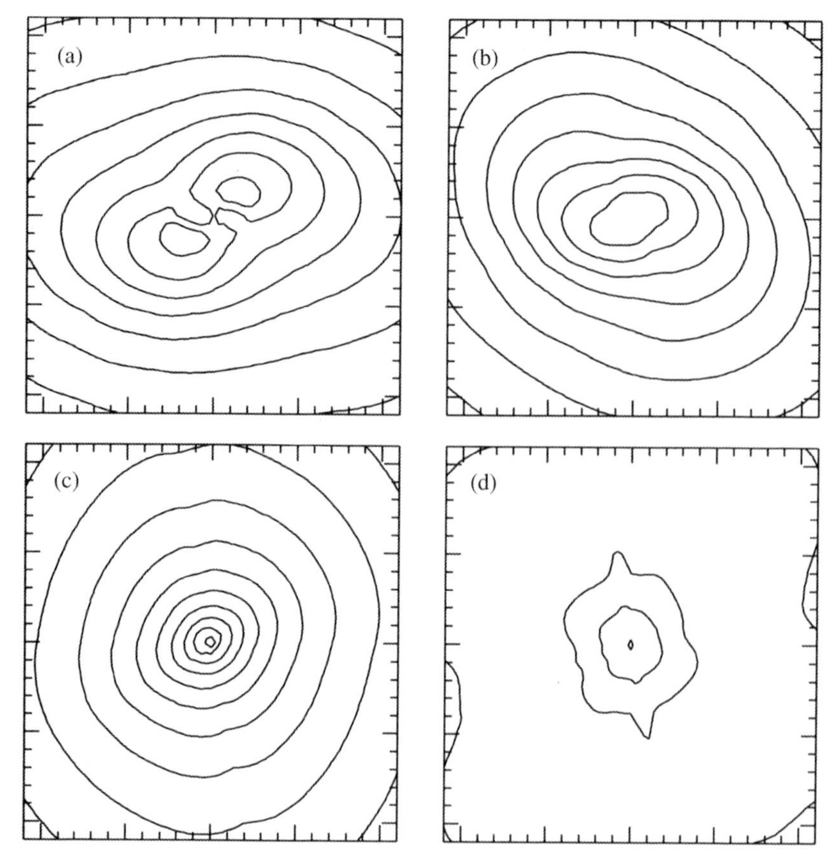

Figure 1.9. Fragmentation of a cloud in fast rotation. Figures a, b and c show the result of numerical simulations of clouds with different degrees of rotation, at about 10 collapse times $t_{collapse}$. Contours represent density in the equatorial plane and increase by a factor 2 from one to the next. The magnetic field is taken into account. For cloud (a) the rotational energy is 0.012 times the gravitational energy; it fragments as a binary star. Cloud (b) for which this ratio is 0.008 only forms a bar, while the disk formed by cloud (c) with a ratio of 0.00012 is almost axi-symmetric. (d) represents the temperature of the gas for cloud (c): it is 25 K, for an initial temperature of 10 K. The collapse then is becoming adiabatic. *From Boss, A.P. (1999) Astrophysical Journal 520, 744, with permission of the American Astronomical Society (AAS).*

Nevertheless, all these models produce stars with too much rotation compared to observations. This irritating problem has recently received an unexpected solution. Observations have shown to everybody's surprise that symmetrical jets originate from stars in formation, including multiple stars (Figure 1.10). This ejection of matter is perpendicular to

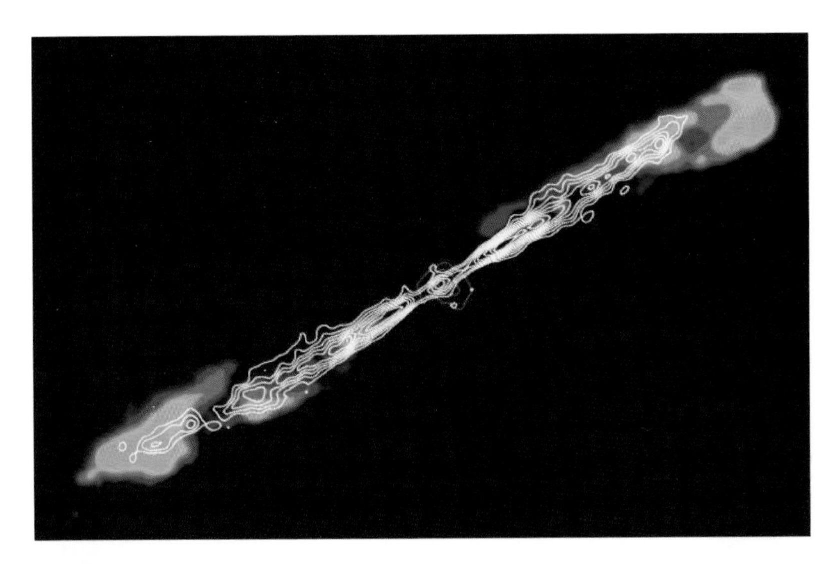

Figure 1.10. The forming star HH211 and its bipolar jet, observed with the interferometer of the French–German–Spanish Institute of Radio Astronomy at Millimetre wavelengths (IRAM). The red contours represent the 1.3-mm emission of dust in the opaque central condensation, not resolved by the instrument. The protostar is inside this condensation. The white contours correspond to the emission of the CO molecule at 2.6 mm, which is contained in the bipolar jet from the protostar. The interaction of this jet with the remains of the parent cloud excites the H_2 molecule, producing its infrared emission (in blue–green). *From Gueth F. & Guilloteau, S. (1999) Astronomy & Astrophysics 343, 571, with permission of ESO and IRAM.*

the protostellar disk, and appears to be a constant of star formation, whatever the mass of the star. The jets have a velocity of several hundred km/s and are made of ionised gas with a temperature of about 10 000 K. They reach several parsecs from the forming star and carry with them the cold surrounding gas: they are surrounded by an envelope of molecular gas whose velocity is smaller, a few tens of km/s. These jets are able to destroy what remains of the parent molecular cloud, but some matter is carried away, and can subsequenty fall on the disk and on the star in formation.

One begins to understand how these jets form. The loss of rotational kinetic energy occurs via a subtle mechanism which involves the magnetic field of the circumstellar disk to pump its rotational energy and to propel the bipolar jet along the rotation axis. But the complexity of this mechanism is such that it is still very difficult to make a global numerical model of the formation of a star.

1.2.3 The different stages of star formation

Figure 1.11 summarises the different steps in star formation. Four successive phases can be distinguished.

The parent cloud is schematised at the top right of this figure, with its dense core and its envelope where density decreases to the exterior. Left, the photon spectrum emitted by this cloud. It is essentially the radiation of dust grains, which are at a very low temperature of the order of 10 K, and takes place in the far infrared and radio waves, with a maximum around 300 μm. This weak radiation has only been observed recently from space, first by balloons and then with the HERSCHEL satellite. There are also molecular lines, not represented on the figure, which are also observed at millimetre and submillimeter waves.

Class 0 objects, below, are collapsing protostars. When a core surrounded by a disk is formed, a bipolar jet appears. The object is still surrounded by a very cold cocoon, and what we observe is mainly the radiation of the cocoon in the far infrared, not different from what was observed before. It might be, however, that millimetre-subillimeter interferometers reach in the near future a sufficient sensitivity and resolving power to observe in some cases what occurs inside. The symmetrical jets are already well observed in radio and infrared light, but we only see in visible light the interaction zones between these jets and the surrounding medium: these are the *Herbig–Haro objects*, from the name of those who described them first.

The next step corresponds to Class I objects. Now the bipolar jet has sufficiently dispersed the molecular cloud which surrounds the protostar, so that one begins to see this protostar in the near infrared (spectrum in grey), as well as the disk which radiates in the mean infrared. The remnants of the molecular cloud are heated by the protostar, so that their radiation is at shorter wavelengths than before, with a maximum at 50–100 μm.

For Class II, the molecular cloud is entirely dispersed and the central star and its surrounding disk appear in full light. The Hubble Space Telescope has allowed us to see numerous objects in this phase, which can still exhibit a bipolar jet, but fainter than in the preceding phases (Figure 1.12). The radiation of the star and of the dust in the disk are no

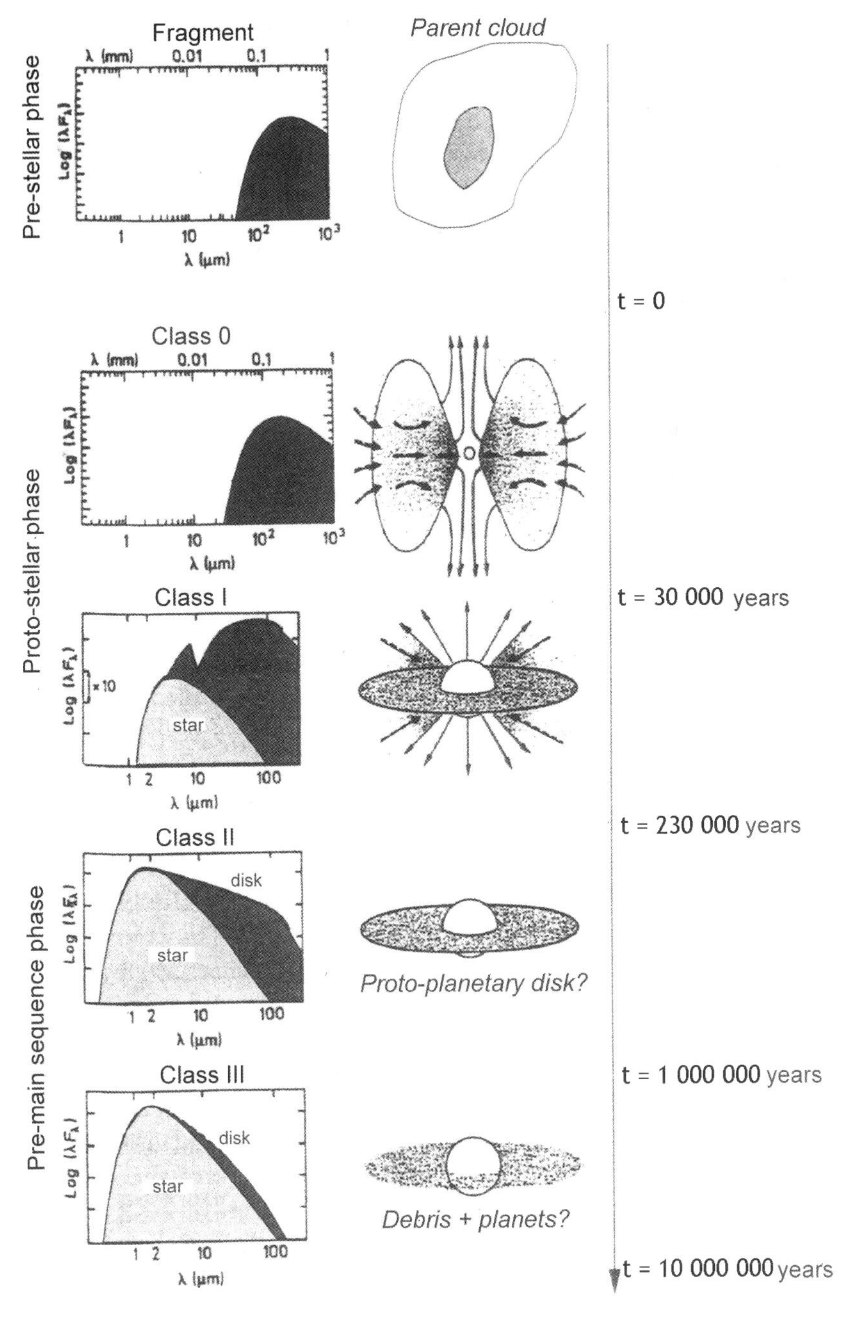

Figure 1.11. The different stages of star formation. *From Philippe André and collaborators.*

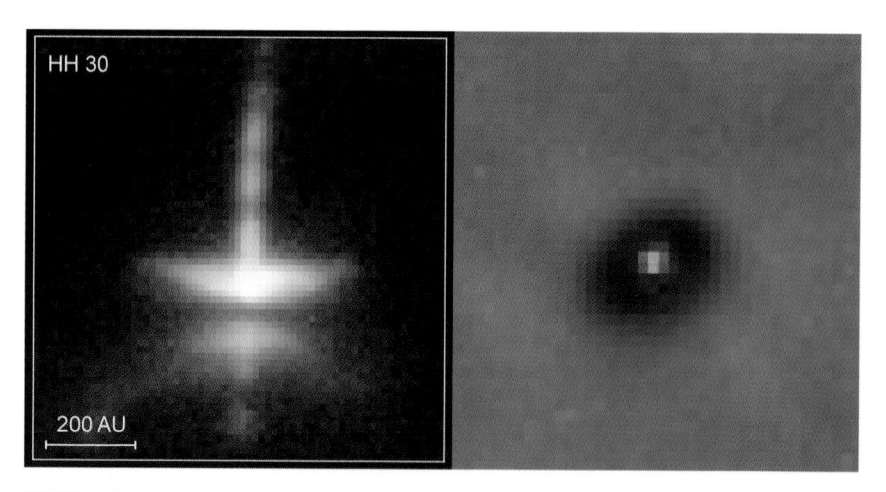

Figure 1.12. Two protostellar disks seen with the Hubble Space Telescope. Right, one sees the protostar at the middle of a disk seen nearly face on; the dust of the disk absorbs the light of the luminous background. Left, the system is seen edge on, so that the central protostar is hidden by the dust in the disk which seems to cut the image into two halves: but the protostar illuminates the visible parts of the disk. One also sees the symmetrical jets originating in the protostar. © *Hubble Space Telescope Gallery.*

longer affected by absorption in the molecular cloud and appear as they are. Observations show that the disk has a diameter of some 500 astronomical units, and is therefore much more extended than the Solar System. Can one speak of a protoplanetary disk in these conditions? This is dubious. The stars in this phase are called *T Tauri stars*, from the name of their prototype.

For massive stars, these different steps are very difficult to observe, because the protostars and disks are buried for a long time in a cocoon that is completely opaque to visible and even infrared radiation. However, it has recently been possible to observe some protostellar disks in the mean infrared. An example is shown in Figure 1.13.

When the accretion disk has been almost entirely captured by the star, its angular momentum being dissipated through viscosity and jets, these jets disappear and there is little matter left around the star. This corresponds to Class III objects. They are now surrounded by a cold disk containing solid fragments resulting from the agglomeration of dust grains. These are the rocks from which planets will form. When they are far from

Figure 1.13. A massive protostellar disk observed at the wavelength of 10 μm with one of the 8-m telescopes of the ESO VLT. The central star, not visible on this image, is probably of type O9, with a mass of about 20 solar masses. © *ESO.*

the star they can capture a fraction of the residual gas: this was the case for the giant planets of the Solar System. Still further away, the solid nuclei can be covered by ice, forming iced objects like Pluto and the comets, which therefore represent the material of the primitive Solar System.

1.3 Stellar evolution before the main sequence

1.3.1 The formation of stars, properly speaking

Let us come back to the central condensation of a collapsing cloud, which is going to form a star. When it becomes opaque to radiation, it contracts slowly, and its gravitational energy progressively transforms into heat and radiation. As one is near equilibrium, the virial equation (1.2) is valid. The roles of external pressure and the magnetic field are now negligible, so that there is no need to use Equation (1.5). Rotation has been braked, so that the rotational kinetic energy is small with respect to the kinetic thermal energy E_{th}. The kinetic energy E_{kin} is therefore reduced to E_{th}. One may thus write the virial equation as:

$$2E_{th} = -E_{pot} = 3/5\ GM^2/R, \tag{1.15}$$

assuming for simplification the density to be uniform, a relatively correct assumption at the beginning of the contraction. During contraction, one must take into account that for a moving particle $E_{kin} + E_{pot} = 0$. Therefore, *half of the potential energy must be radiated during contraction, the other half being transformed into thermal energy.*

The time t_{KH} for contraction to some radius R_{final} can be evaluated, assuming the luminosity L of the star to be constant:

$$t_{KH} = 1/2\ L[E_{pot}(R_{initial}) - E_{pot}(R_{final})] \approx -1/2\ LE_{pot}(R_{final})$$

$$= 3/10\ GM^2/LR_{final}. \tag{1.16}$$

t_{KH} is the Kelvin–Helmholtz time, from the name of the two physicists who made this calculation first. For $M = 1\ M_\odot$, this time is only of the order of 10^7 years. This shows that gravitational energy cannot be the sole source of luminosity for the Sun and the stars, as it was still believed at the end of the 19th century.

When contraction becomes fully adiabatic, the thermal energy cannot be radiated efficiently, and the relation between the temperature T and the specific mass ρ is $T \propto \rho^{\gamma-1}$, with $\gamma = C_P/C_V$, the ratio of specific heats respectively at constant pressure and at constant volume. For the Jeans mass $M_{Jeans} \propto T^{3/2}\rho^{-1/2}$ (equation 1.6), $T \propto \rho^{1/3}$ for M_{Jeans} = constant. This means that a Jeans mass contracts slowly only if, during contraction, $\gamma-1$ becomes larger than 1/3. If not, the mass is unstable and a quasi-equilibrium cannot take place.[3] This is the case as long as hydrogen is neutral and opacity not extremely large: then the mass collapses as seen earlier. But γ grows when ionisation occurs due to the rise of temperature, and becomes larger than 4/3: for a fully ionised medium, $\gamma = 5/3$. Then temperature

[4] This property is also valid if there is an external pressure, see equation (1.11).

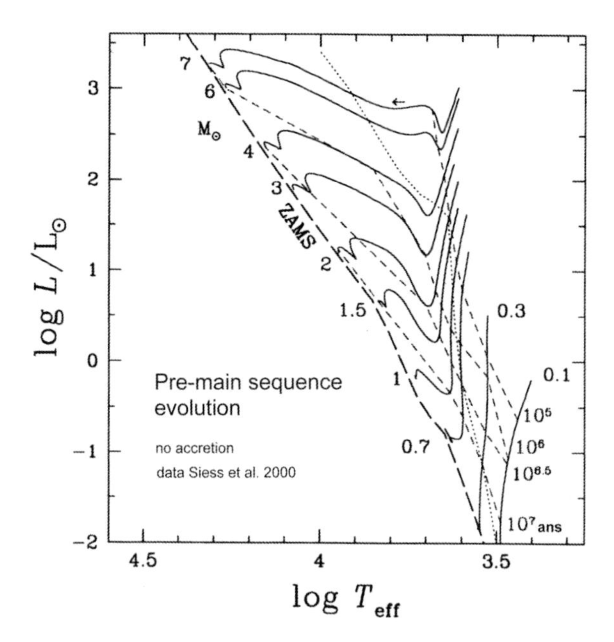

Figure 1.14. The evolution of stars before the main sequence. The full lines are the evolution tracks of young star of different masses until the zero age main sequence (ZAMS). The dashed lines are isochrones for dates indicated at the bottom right. The dotted line corresponds to the beginning of hydrogen fusion into helium. If the stars accrete external matter, which is the most probable case, the evolution follows rather different tracks. *From de Boer & Seggewiss (2008).*

rises further and one reaches an equilibrium radius. Actually, the medium becomes very opaque at ionisation, due to the Thomson scattering of photons on free electrons.[4]

During this phase, the star is purely convective except in its central region, because heat can only be carried to the exterior by convection. The surface temperature stays approximately constant, of the order of 4 000 K, the ionisation temperature of hydrogen. However, since the star radiates it continues to contract, but much more slowly than before: consequently, its luminosity decreases slowly at a roughly constant temperature. The corresponding tracks in a luminosity vs. temperature diagram are called *Hayashi tracks*, from the name of the astronomer who produced the first

[5] It is for the same reason that the Universe is opaque at redshifts larger than 1 000, for which the temperature is larger than 3 000 K, enough for partial ionisation of hydrogen.

important work on this evolution (Figure 1.14). If the star continues to accrete matter, the evolution is rather different. If the mass is sufficient, the central temperature and density reach values for which hydrogen fusion into helium is possible. Then the star experiences internal adjustments until it stabilises on the main sequence.

Stars in this pre-main sequence phase are the T Tauri stars. The large-mass ones are the Ae and Be stars. The "e" suffix means that their spectrum, as indeed that of the T Tauri, exhibits emission lines which come from the protostellar disk, not from the star itself. All these stars are very active: they are variable and emit X rays, very probably due to the magnetic activity at their surfaces.

1.3.2 The brown dwarfs, aborted stars

Figure 1.15 shows the time evolution of the central temperature of protostars with different masses. A protostar must have a mass larger than 0.07 $M_\odot$ for hydrogen fusion to occur, becoming then a normal star. However, lithium which is present in small quantities in the interstellar medium (see Table 1.1) can burn at a lower temperature than hydrogen, and thus in stars with slightly smaller masses. Therefore, it is fully destroyed in normal stars. The same is true for deuterium which burns in stars with masses as small as 0.015 $M_\odot$. These objects therefore possess a temporary source of nuclear energy which impedes for a few million years their slow, but ineluctable cooling. Their surface temperature is lower than 2 500 K, so that they radiate mainly in the red and infrared, hence their name of brown dwarfs. Their spectrum is characterised by bands of various molecules like H_2, CO, CH_4, NH_3, H_2O, etc.

Figure 1.16 shows the first brown dwarf discovered, in 1994, close to a dwarf star with which it forms a gravitationally tied binary system. One knows today a fairly large number of brown dwarfs, in particular in the Pleiades. From their number and a theoretical study of their cooling, one can derive their initial mass function, and calculate that their total mass is of the order of 10% of the total mass of stars in our Galaxy.

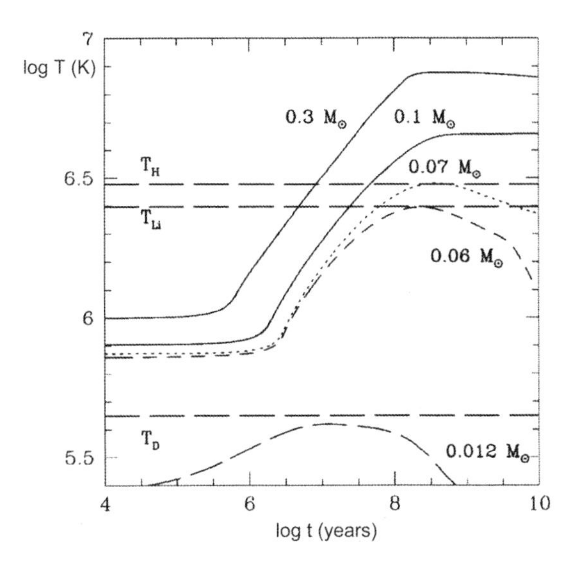

Figure 1.15. Time evolution of the central temperature of protostars of various masses. The temperatures for which fusion of deuterium (T_D), lithium (T_{Li}) and hydrogen (T_H) are indicated. One sees that only protostars with masses larger than 0.07 M⊙ can reach the fusion temperature of hydrogen. *From Chabrier, G. & Baraffe, I. (2000) Annual Review of Astronomy & Astrophysics 38, 337.*

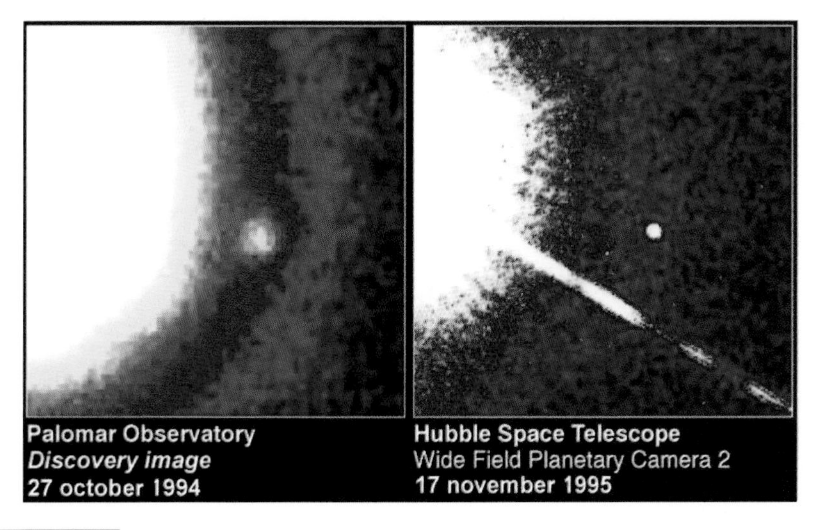

Figure 1.16. The brown dwarf Gliese 229B next to the normal dwarf star Gliese 229. It is very weak and could only be studied in detail with the Hubble Space Telescope. The line originating from the main star in the figure at the right is an artefact due to diffraction by the support of the secondary mirror of the telescope. © *Hubble Space Telescope Gallery.*

2

The Physics of Stars

2.1 The fundamental parameters: mass, radius, luminosity; the Hertzprung–Russell diagram

The spectral distribution of the radiation of stars does not differ much from that of a blackbody. As a consequence, we generally express the luminosity L of the star, i.e. the total radiated power, as an *effective temperature* T_{eff}, which is the temperature that a fictitious blackbody with the same dimensions as the star and emitting the same power would have. We have the following relation between L, T_{eff} and the radius R of the star:

$$L = 4\pi R^2 \sigma T_{eff}^4, \tag{2.1}$$

σ being the Stefan–Boltzmann constant. However, it is difficult to obtain directly these quantities. A precise determination of the effective temperature would require a quantitative measurement of the spectral distribution of the stellar radiation from the far ultraviolet to the mean infrared, which necessitates observations from the ground well corrected for atmospheric extinction as well as observations from space vehicles. We would measure at the same time the *apparent luminosity* of the star, i.e. the power received from it per unit area above the atmosphere. But we have to know its distance to determine its *absolute*

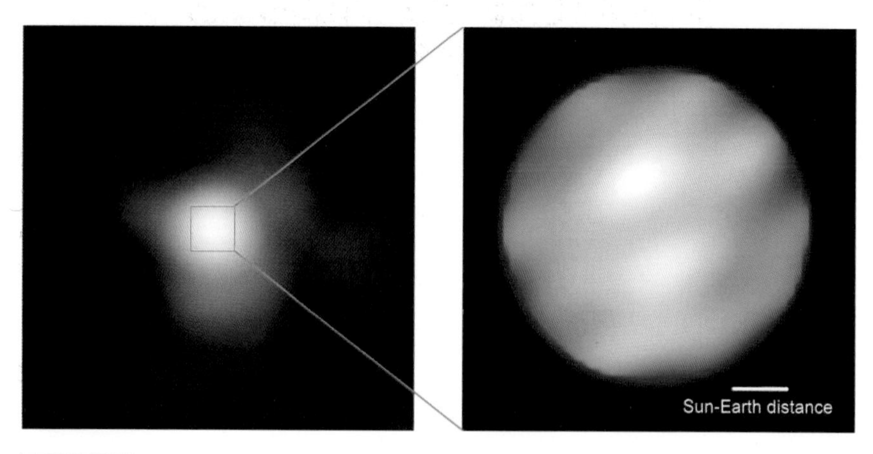

Figure 2.1. Infrared images of the red supergiant Betelgeuse = α Orioni (spectral type M2Iab). Left, an image obtained with adaptive optics adapted to one of the 8-m telescopes of the VLT of ESO (© *ESO and P. Kervella*); the angular resolution of 0.037 arc second is barely sufficient to resolve the star itself but the matter it ejects is visible. Right, an image obtained by the Infrared Optical Telescope Array (IOTA) interferometer in Arizona, with an angular resolution of 0.009 arc second (*From Haubois, X. et al. (2009) Astronomy & Astrophysics, 508, 923–932, with permission of ESO*). The angular diameter of the star is 0.045 arc second and its average effective temperature is 3 600 K but there are large variations over the surface of the star reaching ±500 K.

luminosity L. It is possible to measure directly the angular size of the star, either by studying the eclipses of binary stars by one another, or directly by interferometry. The first such systematic interferometric measurements were made by Hanbury Brown and Twiss in 1956. Presently, these measures have become routine, and we can even obtain real images of nearby giant stars like Betelgeuse (Figure 2.1). From the angular radius, we can derive the effective temperature and the apparent luminosity without knowing the distance to the star, using an equation similar to (2.1).

Observations of stars of different types allowed us to obtain a relation between their colour in the visible domain and their effective temperature, so that it is sufficient to measure the colour to derive an approximate value of T_{eff}. This colour is defined from flux measurements through coloured filters. There are several such photometric systems, but we will only consider here that of Johnson.

Filter	Mean λ (nm)	Width $\Delta\lambda$ (nm)	e_0 (erg s^{-1} cm^{-2} nm^{-1})	
U	360	68	4.35×10^{-8}	Ultraviolet
B	440	98	7.20×10^{-8}	Blue
V	550	89	3.92×10^{-8}	Visible

Table 2.1. Parameters of the filters UBV of Johnson.

The fluxes are generally expressed in a logarithmic scale of *magnitudes*, a quantitative version of the old qualitative system. If $e(\lambda)$ is the monochromatic irradiance due to the star above the terrestrial atmosphere, the corresponding magnitude at the wavelength λ is:

$$m(\lambda) = -2.5 \log e(\lambda)/e_0(\lambda), \tag{2.2}$$

where the constant $e_0(\lambda)$ defines the zero magnitude. In practice, the measurement is performed in a more or less wide spectral band defined by a filter, and equation (2.2) must be integrated on this band. Table 2.1 gives the parameters of the three filters U, B and V of Johnson:

Table 2.2 gives for stars of different spectral types the B–V colour (B and V designating now the magnitudes in the corresponding filters), the effective temperature and the *absolute magnitude* M_V in filter V. By definition, the absolute magnitude is the apparent magnitude that the star would have above the atmosphere at a distance of 10 parsecs. We have therefore the following relation between the apparent magnitude m and the absolute magnitude M:

$$M = m + 5 + 5 \log D, \tag{2.3}$$

D being the distance in parsecs. In particular, this relation applies for magnitudes M_V and V.

In order to obtain the radius of the star and its absolute luminosity, we have to know its distance, which can be determined in three different ways:

- by measuring its *geometric parallax*. A star describes every year an ellipse in the sky with respect to very distant objects like galaxies or

Type	Dwarfs (type V)				Giants (Type III)				Supergiants (Type I)			
	$B-V$	T_{eff} (K)	M_V	$L(L_\odot)$	$B-V$	T_{eff} (K)	M_V	$L(L_\odot)$	$B-V$	T_{eff} (K)	M_V	$L(L_\odot)$
WN									-	≈ 40 000	−8.0	$\approx 10^5$
WC									-	≈ 50 000	−8.5	$\approx 2 \times 10^5$
O5	−0.33	44 500	−5.7	4.2×10^5					−0.30	40 300	−6.6	1.1×10^6
B0	−0.30	30 000	−4.0	5.2×10^4					−0.25	26 000	−6.4	2.6×10^5
B5	−0.17	15 200	−1.2	8.3×10^2					−0.10	13 600	−6.2	5.2×10^4
A0	−0.02	9 520	0.6	54	−0.03	10 100	0.0	106	−0.01	9 730	−6.3	3.5×10^4
A5	0.15	8 200	1.9	14	0.15	8 100	0.7	43	0.09	8 510	−6.6	3.5×10^4
F0	0.30	7 200	2.7	6.5	0.30	7 150	1.5	20	0.17	7 700	−6.6	3.2×10^4
F5	0.44	6 440	3.5	3.2	0.43	6 470	1.6	17	0.32	6 900	−6.6	3.2×10^4
G0	0.58	6 030	4.4	1.5	0.65	5 850	1.0	34	0.76	5 550	−6.4	3.0×10^4
G5	0.68	5 770	5.1	0.79	0.86	5 150	0.9	43	1.02	4 850	−6.2	2.9×10^4
K0	0.81	5 250	5.9	0.42	1.00	4 750	0.7	60	1.25	4 420	−6.0	2.9×10^4
K5	1.15	4 350	7.4	0.15	1.50	3 950	− 0.2	170	1.60	3 850	−5.8	3.8×10^4
M0	1.40	3 850	8.8	7.7×10^{-2}	1.56	3 800	− 0.4	330	1.67	3 650	−5.6	4.1×10^4
M5	1.64	3 240	12.3	1.1×10^{-2}	1.63	3 330	− 0.3	930	1.80	2 800	−5.6	3.0×10^5

Table 2.2. Parameters of stars of different types.

quasars, because of the parallax effect due to the orbital motion of the Earth. The semi-major axis of this ellipse, which is the parallax π, is equal to 1 second of degree for a star at a distance of 1 parsec. The European satellite HIPPARCOS has allowed us to measure geometric parallaxes of a very large number of stars with an accuracy of 0.001 or somewhat better: the distance of these stars is therefore known to an accuracy better than 10% up to a distance of 100 pc. The satellite GAIA will do even better;
- by determining the average of the *proper motions* in the sky of stars which are considered to belong to a common group (*statistical parallax*). The proper motion μ, expressed in seconds of degree per year, is then oriented in a direction opposite to the motion of the Sun with respect to the nearby stars, a motion which has a velocity of 20 km s^{-1} towards an apex located in the direction of the constellation Cygnus. The proper motion is of course smaller for larger distances of the observed stars. For stars located at 42 pc in a direction perpendicular to that of the apex, the proper motion is 0.1″/year, which is easily measurable. This method does not work for isolated stars, which have too high random motions;
- by comparing the apparent magnitudes of two stars which exhibit similar characteristics, and are therefore considered as identical; the distance of the nearer one being known by one of the preceding methods, we can obtain easily that of the more distant, after correcting for the effects of interstellar extinction (*photometric parallax*). It is in this way that Henrietta Lewitt obtained in 1912 the distance of the Small Magellanic Cloud, a satellite galaxy of our own Galaxy, by comparing the magnitudes of cepheids, a particular class of variable stars, to that of similar stars in the Galaxy.

If we plot as a function of colour $B–V$ the absolute magnitude M_V of stars with known distances, we obtain the *Hertzprung–Russell diagram* (HR diagram), from the name of the two astronomers who popularised the first such diagrams. Figure 2.2 shows two versions of the HR diagram for nearby stars. We notice that the representative points of the different stars are not distributed at random, but that most of them, the *dwarfs*, are grouped around a line called the *main sequence*, where stars stay during approximately nine

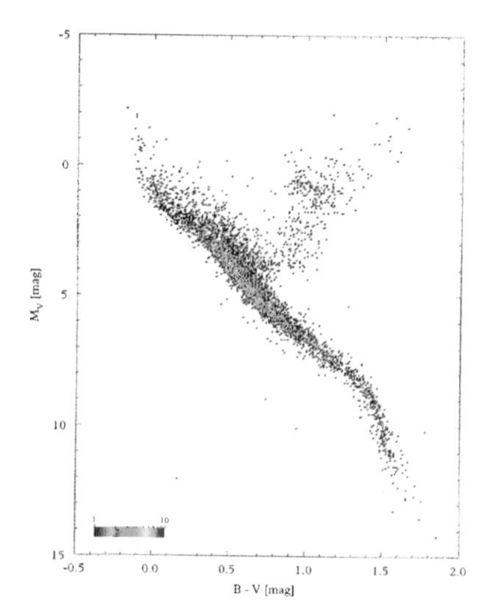

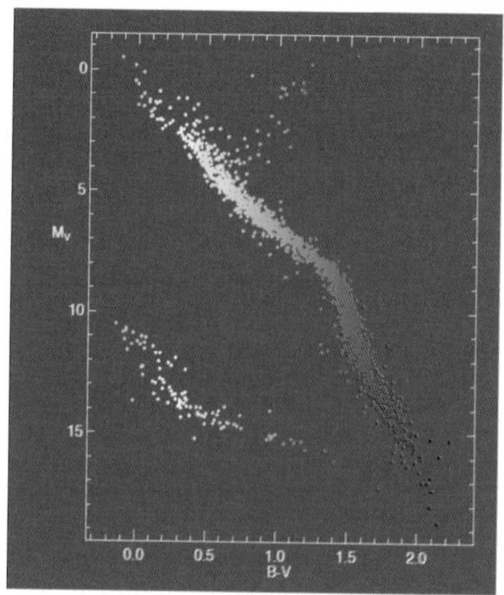

Figure 2.2. Left, the Hertzprung–Russell diagram (HR diagram) obtained with the HIPPARCOS satellite for the brightest stars of the sky: © *ESA*. It contains 4 907 stars whose distance is known to better than 5%. The colours indicate the representative points for which there are more than one star. To identify the different types of stars see Fig. 2.3. Right, the HR diagram for 1090 stars contained in a small volume around the Sun: *from Jahreiss, H. & Gliese, W. (1993) IAU Symposium 156, 107.* It gives a better idea of the proportions of the different types of stars. Notice in particular the small number of giants and the large quantity of white dwarfs.

tenths of their lives. Another branch starts from the main sequence: here, the luminosity increases with *B–V*, hence for redder thus cooler stars. What differentiates these stars from the main sequence ones of the same colour, hence of similar effective temperatures, is their higher luminosity and hence their larger radius: these are the *red giants*, which are in a more rapid phase of evolution than the dwarfs. Figure 2.3 allows one to identify the different types of stars in the HR diagram, and Figure 2.4 shows the HR diagram of stars in a cluster aged 4 billion years, M 67.

The *B–V*, M_V diagram can be transformed into a diagram where L is plotted as a function of T_{eff}. We obtain in this way the *theoretical HR diagram*, which can be easily compared to the predictions of models of stellar evolution.

The mass of stars can only be determined accurately by observation of eclipsing binary stars, for which one of the components passes

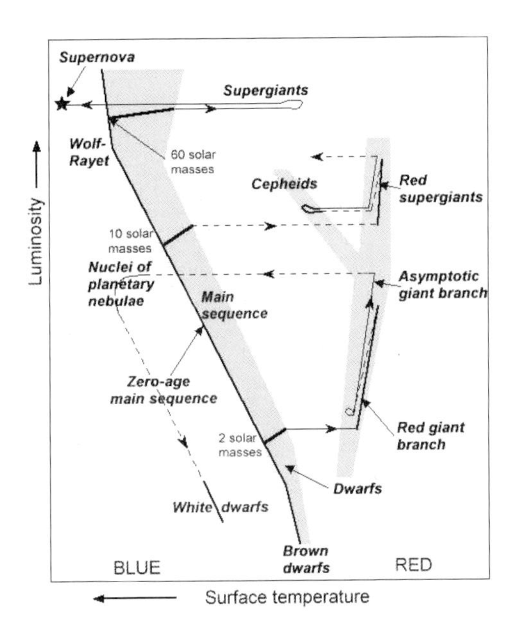

Figure 2.3. A Hertzprung–Russell diagram, where the representative zones for different types of stars and their evolutionary tracks are indicated in a schematic way.

alternatively in front and behind the other, producing variations in the light received from the system: in this case, the Earth is approximately in the plane of the orbit. After measuring the radial velocity of the components by spectroscopy using the Doppler–Fizeau effect, we apply Kepler's laws to obtain the masses of the individual stars.[1] The comparison of the masses obtained in this way with luminosities give a remarkable result: for main sequence stars, mass is closely related to luminosity (Figure 2.5). The more massive the star, the more luminous it is. We derive from this an important property: the lifetime on the main sequence is shorter for more massive stars. Indeed, this lifetime goes like the ratio of the mass (the reservoir of energy) with luminosity (the rate of consumption of this energy). The lifetime of the Sun is of the order of 10 billion years, but that of a 10 $M_\odot$ star, whose luminosity is about 10 000 times higher, is only 10 million years.

[1] We come back to this point in detail in Chapter 5.

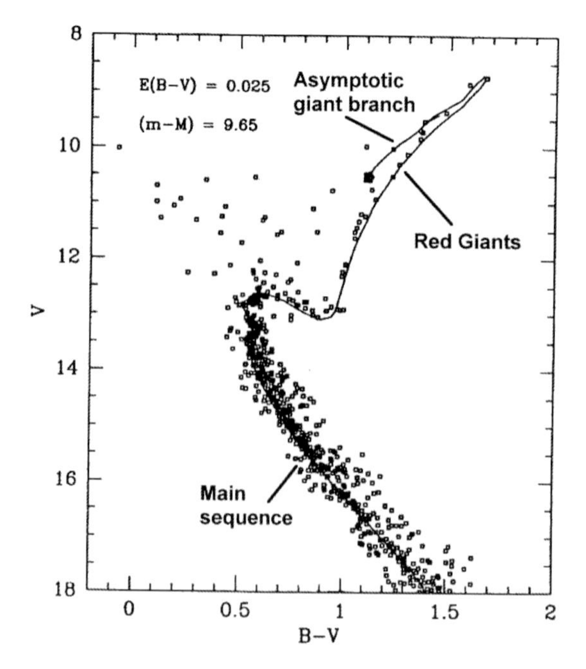

Figure 2.4. The HR diagram of the star cluster M 67. All the stars in this cluster have the same age of 4 billion years, and their abundances in heavy elements are close to that in the Sun. The line represents the theoretical locus where stars of this age are found. Stars of the red giant branch and the asymptotic giant branch have almost all the same mass, slightly larger than that of the Sun, because their evolution is fast. The stars which are outside the line are probably double stars. *From Carraro, G. et al. (1996) Astronomy & Astrophysics 305, 849, with permission of ESO.*

2.2 Stellar atmospheres

The visible spectrum of most stars consists of a continuum with atomic absorption lines (also molecular absorption lines for the coldest ones). Wolf–Rayet stars, which have emission lines, are a remarkable exception. It has been believed for a long time that the absorption lines were formed in a relatively cold *reversing layer* located above a hotter zone which emitted the continuum, the *photosphere*. This is erroneous: the continuum radiation and the lines are the result of radiation transfer in a thick layer of the atmosphere, and it is not possible to assign a definite depth for their formation. However, because opacity is larger in a line than in the adjacent

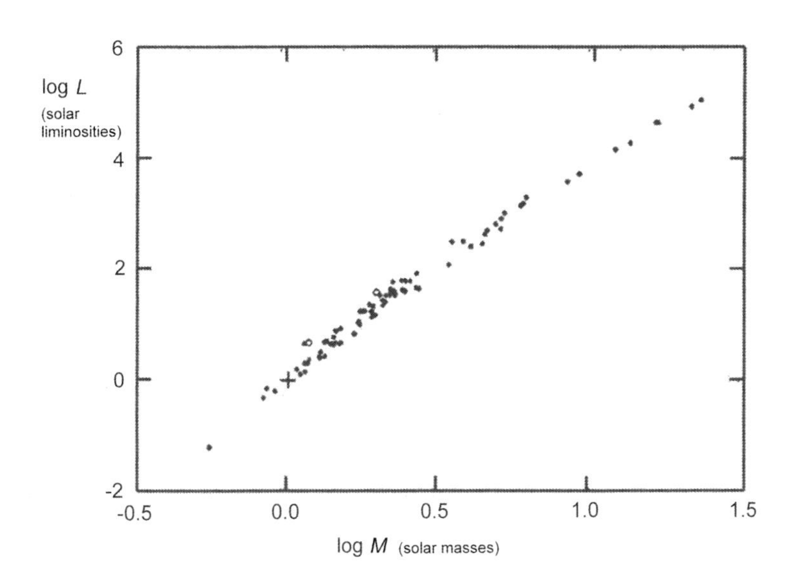

Figure 2.5. The mass–luminosity relation for eclipsing binaries for which spectral lines are observed for each of the two components. Luminosities and masses are in solar units. The position of the Sun is indicated by a cross. *From Zahn, J.-P., in Lequeux et al. (2009).*

continuum, we see in a line shallower zones than in the continuum. In both cases, the emission is close to that of a blackbody at the considered wavelength and at the average depth of the emitting region. If temperature decreases with altitude in the atmosphere, the line is darker than the continuum and appears is absorption lines. This is the reverse in some regions of the stellar atmosphere like the solar chromosphere, where temperature increases with elevation.

We call *photosphere* the zone from which the bulk of the visible light comes, but this is an ill-defined notion since its depth depends on wavelength, and also on the angle of view. In fact, a consequence of the existence of a temperature gradient is the limb darkening of the disk of the Sun or stars: as the atmosphere near the limb is seen obliquely, the opacity in the continuum is such that we only see the radiation of the upper layers, which are colder and thus dimmer than those which we see at the centre of the disk.

The origin of the continuum has taken a long time to identify. For stars with an effective temperature smaller than 6 000 K, the Sun in particular, the continuum is emitted by the ion H^- (a hydrogen atom with an extra electron), as proposed in 1939 by Rupert Wildt, and established in a definitive way in 1946 by Chandrasekhar and Breen, who calculated the absorption corresponding to the ionisation of this ion to form a hydrogen atom: $H^- + h\nu \leftrightarrow H + e^-$. For hotter stars, the continuum is dominated by ionisation of hydrogen, as observed by Jules Baillaud in 1926 at the Pic du Midi observatory: $H + h\nu \leftrightarrow H^+ + e^-$. The Thomson diffusion of light by free electrons dominates the opacity for still hotter stars, where hydrogen is almost fully ionised in their atmospheres. The photodissociation of molecules contributes to the continuum for the coldest stars, and the ionisation of helium to that of the hottest ones.

The spectrum of stars depends much on their temperature. The spectral distribution of the continuum is not very different from that of a black-body at temperature T_{eff}, so that the hot stars have a maximum of emission in the ultraviolet (in the far ultraviolet for O stars, which is observable only from space vehicles), and the cold stars emit mainly in the red or the infrared. The lines which dominate the spectrum of the hot stars are those of hydrogen, and even those of helium for the hottest ones. Metallic lines appear more and more numerously in colder and colder stars, then molecular bands (Figure 2.6). This gives the basis of stellar classification, which can be briefly summarised as follows:

WR : broad emission lines
O : absorption lines of hydrogen, neutral and ionised helium;
B : absorption lines of hydrogen and neutral helium;
A : hydrogen lines dominate;
F : many lines of ionised metals;
G : lines of ionised and neutral metals;
K : lines of neutral metals dominate;
M : molecular bands, some of oxygenated molecules like TiO;
C : bands of carbonaceous molecules;
S : bands of zirconium oxide ZrO.

The WR (Wolf–Rayet, from the name of their discoverers), C and S stars are rare evolved stars which do not belong to the main sequence.

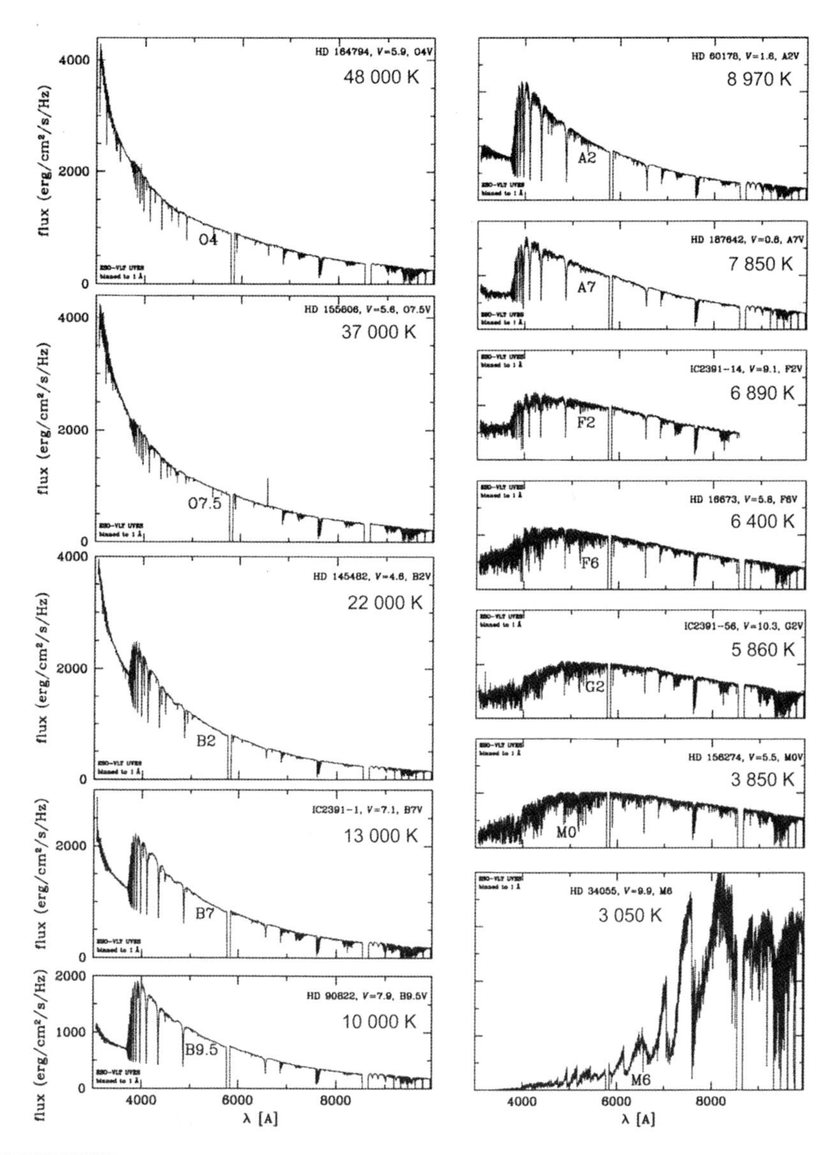

Figure 2.6. Spectra of main sequence stars. The name of the star, its visual magnitude *V* and its spectral type are indicated in each box, as well as its effective temperature. The spectra are made of three parts which leave small empty intervals near 5 750 and 8 600 Å. They are smoothed to a resolution of 1 Å. Note the progressive change in the spectral energy distribution with temperature, and also that the number of absorption lines increases when temperature diminishes. The Balmer discontinuity of hydrogen at 3 840 Å is conspicuous for B and A stars. *Data from the UVES spectrograph at the Very Large Telescope of ESO, from de Boer & Seggewiss (2008).*

For main-sequence A to G stars, in particular the Sun which is a G5 star, the temperature in the atmosphere, after going through a minimum of about 4 000 K, increases with altitude. This produces emission lines which are not easy to see, except in the far ultraviolet or in the visible, by high-resolution spectroscopy at the centre of some absorption lines which are produced deeper in the atmosphere. The corresponding layer is called the *chromosphere*, a name which comes from the fact that, when we can see it around the Sun during total eclipses, it is pink in colour due to the strong Hα emission line of hydrogen at the wavelength of 656.3 nm. Above the chromosphere, the gas is hotter and hotter while being more tenuous, and reaches nearly a million K: this is the *corona*, which is visible around the Sun during total eclipses, or with specialised instruments, the coronographs. The corona emits lines of multi-ionised atoms and thermal X rays. Both the chromosphere and the corona appear to be heated by magnetic processes.

Many stars produce winds. The Sun is an example, its wind being due in this case to the evaporation of the corona and to more or less localised ejections of matter. This phenomenon is normal and very important for very hot stars: they lose in this way an important fraction of their mass during their evolution, which is strongly affected by this loss as we will see in the next chapter, Section 3.3. The velocity of the wind can reach several thousand kilometres per second. Stars colder than about 2 500 K also emit a wind, but the mechanism is different from that for hot stars: here, dust grains condense in the atmosphere, and are pushed outwards by radiation pressure, driving the gas through viscosity with a velocity of a few tens of km/s.

We will stop here our short presentation of stellar atmospheres. The theory of these atmospheres, and its use to derive the abundances of elements, is a complex and fastidious affair that we consider as outside the scope of this book.

2.3 The basic equations of stellar structure

The structure of a star is defined by four differential equations that we will examine in succession. They assume spherical symmetry and also that the star is made of an approximately perfect gas.

2.3.1 The equation of mass continuity

A spherical shell of thickness dr at radius r contains the mass $dM(r) = 4\pi r^2 \rho(r) dr$, where $\rho(r)$ is the specific mass per unit volume. We deduce immediately:

$$dM(r)/dr = 4\pi r^2 \rho(r) \qquad (2.4)$$

2.3.2 The equation of hydrostatic equilibrium

The force of gravitation is balanced everywhere by the pressure gradient, with its different components, including the radiation pressure. The gravitational force per unit area on a layer of thickness dr is:

$$dF = -GM_r \rho(r)/r^2 \, dr, \qquad (2.5)$$

where M_r is the mass inside the radius r. The pressure difference which equilibrates gravitation is $dP(r)$. We can write:

$$dP(r)/dr = -GM_r \rho(r)/r^2 \qquad (2.6)$$

2.3.3 The equation of energy conservation

Let us assume that all the energy produced inside the star is carried to the surface and radiated. Let $L(r)$ be the total flux entering a spherical shell of radius r, and $dL(r)$ the additional flux which might be generated in this shell. If $\varepsilon(r)$ is the energy generation rate per unit mass at this radius, we can write the energy conservation equation as:

$$dL(r)/dr = 4\pi r^2 \rho(r) \varepsilon(r) \qquad (2.7)$$

However, if there is no equilibrium, there might be an additional gain or loss of energy, e.g. as the work of contraction or expansion. We can show that the energy conservation equation becomes:

$$dL(r)/dr = 4\pi r^2 \rho(r) \{\varepsilon(r) - C_p dT(r)/dt + [1/\rho(r)] dP(r)/dt\}, \qquad (2.8)$$

C_p being the specific heat at constant pressure, $T(r)$ the temperature, $P(r)$ the pressure and t the time.

2.3.4 The equation of temperature gradient

This is the last fundamental equation. It depends on the way heat is transported: this can be by conduction, radiation or convection. Conduction is generally negligible with respect to the two other modes of transport. The box explains the criteria according to which this transport is by radiation or by convection.

Let us assume first that the transport is through radiation. The equation of radiative transfer is:

$$\cos\theta\, dI_v/dr = -\kappa_v I_v + j_v. \qquad (2.9)$$

I_v is the intensity of radiation at frequency n, which is well approximated by the radiation of a blackbody $B_v \cdot k_v$ is the monochromatic absorption coefficient, and j_v the monochromatic emissivity of the medium. The $\cos\theta$ factor, where θ is the angle of the light ray with the normal to the spherical shell, must be taken into account when considering the transfer in this direction. Let us multiply this equation by $\cos\theta$ and integrate over all directions ϕ around the normal to the shell and over all frequencies ν; we obtain:

$$\iint \cos^2\theta\, (dI_v/dr)\, d\phi\, d\nu = -\iint \cos\theta\, \kappa_v I_v\, d\phi\, d\nu + \iint \cos\theta\, j_v\, d\phi\, d\nu. \qquad (2.10)$$

The integral over ϕ on the left side gives only a factor $4\pi/3$ because only $\cos^2\theta$ depends on the direction. The assimilation of I_v to the blackbody radiation is such that $\int I_v d\nu = \sigma c T^4/4\pi$, where c is the velocity of light. In order to integrate the first term on the right, we can use the flux density F_v which is the integral of I_v over all directions: $F_v = \int I_v \cos\theta\, d\phi$. The integral of the second term on the right is zero because j_v does not depend on direction. We obtain:

$$4\pi/3d(\sigma c T^4/4\pi)/dr = -\rho(r) \int \kappa_v F_v\, d\nu, \qquad (2.11)$$

where the absorption κ_ν is taken per unit mass, in $cm^2\ g^{-1}$. The integral on the right side can be simplified by using an absorption coefficient averaged over frequency, the *Rosseland mean* κ_R: $\int \kappa_\nu F_\nu\, d\nu = \int \kappa_\nu\, d\nu \int F_\nu\, dn = \kappa_R \int \nu F_\nu\, d\nu$. We also have $F = \int F_\nu\, d\nu = L(r)/4\pi r^2$. The final equation is:

$$dT/dr\big|_{\text{radiative}} = -3\kappa_R \rho(r)/4\sigma c T^3\, L(r)/4\pi r^2 \tag{2.12}$$

Let us assume now that the transport is via convection. For this, we consider a bubble of gas at radius r. The convection is such that it is slightly hotter than the surrounding gas: it expends and rises. We may assume as a first approximation that the process as adiabatic, so that there is no exchange of heat Q with the ambient medium, $dQ = 0$.

Since $dQ = C_P\, dT - (1/\rho)\, dP$, we have:

$$dT = (1/\rho C_P)\, dP, \tag{2.13}$$

where C_P is the specific heat at constant pressure.

The law of perfect gases $P = \rho \Re T$ gives the density ρ, which we insert into the previous equation, giving:

$$dT = \Re/C_P\ T/P\ dP, \tag{2.14}$$

which can also be written, introducing the adiabatic exponent $\gamma = C_P/C_V$ and taking into account the relation $C_P = C_V + \Re$:

$$dT = (1 - 1/\gamma)\ T/P\ dP. \tag{2.15}$$

We can now write the adiabatic temperature gradient as:

$$dT/dr\big|_{\text{adiabatic}} = (1 - 1/\gamma)\ T/P\ dP/dr \tag{2.16}$$

If the adiabatic gradient is smaller than the radiative gradient in absolute value, the energy transport is convective: see Equation (E1.1) and its demonstration in Box 1.1.

Box 1.1: The Schwarzschild criterion for convection

The temperature gradient determines if the energy transport is by radiation or by convection. Schwarzchild demonstrated that convection occurs when:

$$|dT/dr|_{\text{adiabatic}} < |dT/dr|_{\text{radiative}}. \tag{E1.1}$$

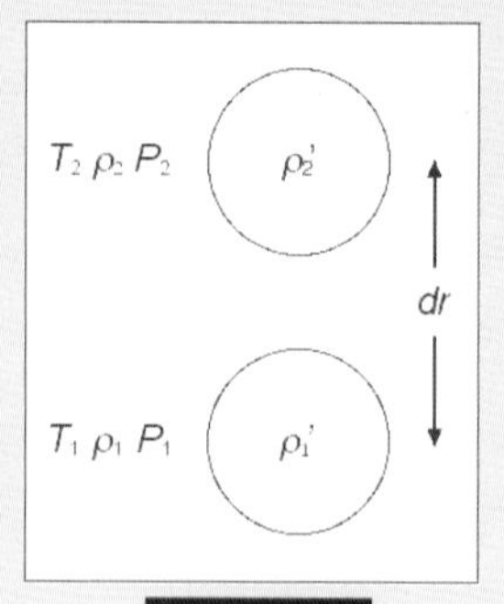

Figure E1.1

To show this, let us consider a bubble of gas immersed in a medium with parameters T_1, ρ_1 and P_1 (Figure E1.1). In general the bubble is approximately in thermal and pressure equilibrium with this medium, but if its specific mass ρ_1' is slightly smaller than ρ_1 it will start to rise. It will then arrive at the level $r + dr$, where the parameters of the ambient medium are T_2, ρ_2 and P_2. It will continue to rise if its new density ρ_2' is smaller than ρ_2. The condition for convection is therefore:

$$\rho_2' < \rho_2, \tag{E1.2}$$

which can also be written:

$$d\rho'/dr < d\rho/dr. \tag{E1.3}$$

The pressures in the bubble and the external medium equilibrate, and this common pressure is $P + dP/dr$. Let us use the law of perfect gases to rewrite (E1.3) as:

$$1/\gamma\ \rho'/P\ dP/dr < \rho/P\ (dP/dr - P/T\ dT/dr), \tag{E1.4}$$

and since $\rho' \approx \rho$,

$$(1 - 1/\gamma)T/P\ dP/dr > dT/dr. \tag{E1.5}$$

We recognise on the left side the adiabatic temperature gradient (Equation 2.16), while the actual temperature gradient of the surrounding gas is on the right side. Because temperature and pressure both decrease towards the surface inside stars, both dP/dr and dT/dr are negative, and the condition for convection is the *Schwarzschild criterion* of (E1.1).

(*Continued*)

Box 1.1 (Continued)

When writing (E1.4), we have implicitly assumed that the average mass per particle μ is a constant. In reality, the chemical composition can change with radius in some zones, and this modifies the criterion for convection. The new criterion is called the *Ledoux criterion*. If it is to be applied, we speak of *thermohaline convection*, by analogy with the situation in terrestrial oceans. If the case is intermediate between those of Schwarzschild and Ledoux, we do not expect convection with the preceding equations, but as convection is not strictly adiabatic, it might nonetheless exist: we speak of this case as *semi-convection*.

2.4 The nuclear reactions, sources of energy for stars

Stars on the main sequence get their energy from the fusion of hydrogen into helium. In later phases of stellar evolution, helium releases energy while transforming into carbon. All these reactions require very high temperatures. We now list them, in order of increasing temperature.

2.4.1 Fusion of hydrogen into helium via the p–p (proton–proton) process; the problem of solar neutrinos

At a temperature larger than 3 million degrees, hydrogen fusion occurs by a series of reactions first described by von Weizsäcker in 1938, in a form less elaborated than here. ^{1}H is the hydrogen nucleus (proton), e^{+} the positron, ν_e the electron neutrino and γ the photon:

$^{1}H + {}^{1}H \rightarrow {}^{2}H + e^{+} + \nu_e$, where ^{2}H is deuterium. This is the slowest reaction of the chain. It is followed by:

$$^{2}H + {}^{1}H \rightarrow {}^{3}He + \gamma .$$

Helium 3 (^{3}He) is transformed into helium 4 (^{4}He) in two principal ways:

$^{3}He + {}^{3}He \rightarrow {}^{4}He + {}^{1}H + {}^{1}H$ (86% of cases), and $^{3}He + {}^{4}He \rightarrow {}^{7}Be + \gamma$ (14% of cases), followed by:

$^{7}Be + e^{-} \rightarrow {}^{7}Li + \nu_e$, then:

$^{7}Li + {}^{1}H \rightarrow {}^{4}He + {}^{4}He$. Be is beryllium and Li is lithium.

There is also a very small probability (0.015% of cases) for ^{7}Be to capture a proton, forming ^{8}B which decays into ^{8}Be*, an excited nucleus which itself disintegrates into 2 ^{4}He:

$$^7\text{Be} + {}^1\text{H} \rightarrow {}^8\text{B} + \gamma\,,$$
$$^8\text{B} \rightarrow {}^8\text{Be}^* + e^+ + \nu_e,$$
$$^8\text{Be}^* \rightarrow {}^4\text{He} + {}^4\text{He}.$$

All this can be summarised by the symbolic reaction:

$$4\,{}^1\text{H} \rightarrow {}^4\text{He} + 2e^+ + 2\nu_e + 24.7\ \text{MeV}.$$

This is what occurs in the central regions of the Sun. The energy production rate of these reactions is:

$$\varepsilon_{p\text{-}p} = 2.5 \times 10^6\, \rho X^2 (T/10^6\ \text{K})^{-2/3} \exp[-33.8/(T/10^6\ \text{K})^{1/3}]\ \text{erg cm}^{-3}\ \text{g}^{-2}\ \text{s}^{-1}, \quad (2.17)$$

where ρ is the specific mass and X is the mass fraction of hydrogen.

Several of the reactions above produce electron-type neutrinos, which can penetrate almost without interaction a large quantity of matter, and thus get out of the Sun and cross the Earth without being attenuated. As early as 1967, the American physicist Raymond Davis attempted to detect the solar neutrinos. He observed a flux one third of that predicted by solar models. This deficit was confirmed by other experiments able to detect electron neutrinos with different energies. The nuclear reaction being well known, there were only two possible explanations for the discrepancy: either the solar models were wrong, or the electron neutrinos disappeared partially on their way to the Earth. Huge theoretical efforts were devoted to solar models, but could only confirm them: there was apparently a deadlock. It took 36 years to solve the problem. However, in 2001, a new neutrino detector was installed in the abandoned mine of Sudbury, in Canada. It was now sensitive to the three species of neutrinos: the electron neutrinos ν_e, the only ones emitted by the Sun, the μ neutrinos and the τ neutrinos. It was then observed that the total number of neutrinos *of the three species* was in agreement with the number of electron neutrinos expected from solar models. The unavoidable conclusion was that the different species of neutrinos could transform into each other, hence the loss of a part of the solar electron neutrinos on their way to the Earth. This necessitated a modification of the standard model of elementary particles, and implied that the neutrinos had a small, but non-zero mass. This idea has been confirmed experimentally in Japan, by studying with a new detector at Kamioka neutrinos emitted by various nuclear reactors within a hundred kilometres. Astrophysics thus allowed an important discovery in physics, for which Davis received the Nobel prize in 2006.

2.4.2 Fusion of hydrogen into helium via the CNO cycles

In massive stars, the central temperature is higher than in the Sun. If it is larger than 10 million degrees, fusion of hydrogen takes place preferably through the following reactions (Bethe, 1938), which are much faster than the p–p reactions: now hydrogen is transformed into helium 4 using pre-existing carbon C, nitrogen N and oxygen O as catalysts (the CNO tri-cycle). A first cycle is:

$$^{12}C + {}^{1}H \rightarrow {}^{13}N + \gamma$$
$$^{13}N \rightarrow {}^{13}C + e^{+} + \nu$$
$$^{13}C + {}^{1}H \rightarrow {}^{14}N + \gamma$$
$$^{14}N + {}^{1}H \rightarrow {}^{15}O + \gamma$$
$$^{15}O \rightarrow {}^{15}N + e^{+} + \nu$$

$^{15}N + {}^{1}H \rightarrow {}^{12}C + {}^{4}He$; another cycle is grafted:

$$^{15}N + {}^{1}H \rightarrow {}^{16}O + \gamma$$
$$^{16}O + {}^{1}H \rightarrow {}^{17}F + \gamma$$
$$^{17}F \rightarrow {}^{17}O + e^{+} + \nu$$

$^{17}O + {}^{1}H \rightarrow {}^{14}N + {}^{4}He$, and also a third cycle:

$$^{17}O + {}^{1}H \rightarrow {}^{18}F + \gamma$$
$$^{18}F \rightarrow {}^{18}O + e^{+} + \nu$$
$$^{18}O + {}^{1}H \rightarrow {}^{15}N + {}^{4}He.$$

The final balance — fusion of four hydrogen nuclei into one helium nucleus — is the same as for the previous process. The energy production rate is now:

$$\varepsilon_{CNO} = 9.5 \times 10^{28}\, \rho X X_{CNO} (T/10^6\ \mathrm{K})^{-2/3} \exp[-152/(T/10^6\ \mathrm{K})^{1/3}]\ \mathrm{erg\ cm^{-3}\ g^{-2}\ s^{-1}}, \tag{2.18}$$

where X_{CNO} is the total fraction of mass as carbon, nitrogen and oxygen.

2.4.3 Helium fusion into carbon (the triple α process)

This fusion occurs at temperatures larger than about 100 million degrees, in the stages of stellar evolution after the sojourn on the main sequence. Two helium nuclei form one beryllium 8 nucleus:

$$^4\mathrm{He} + {}^4\mathrm{He} \leftrightarrow {}^8\mathrm{Be}, \text{ then}$$

$$^8\mathrm{Be} + {}^4\mathrm{He} \rightarrow {}^{12}\mathrm{C}.$$

^8Be is very instable and decays into 2 ^4He in some 10^{-16} s (in other words, the first reaction is reversible). Therefore, the medium must be dense enough so that the beryllium nucleus has the time to meet a helium nucleus before decaying. The reaction rate thus depends on the square of the density and is:

$$\varepsilon_{3\alpha} = 2 \times 10^{17}\, \rho^2 Y^3 (T/10^6\ \mathrm{K})^{-3} \exp[-4670/(T/10^6\ \mathrm{K})]\ \mathrm{erg\ cm^{-3}\ g^{-2}\ s^{-1}}, \tag{2.19}$$

where Y is the mass fraction of helium.

2.5 Modelling the stellar interiors

The simultaneous resolution of the four basic differential equations (2.4), (2.6), (2.7) and (2.12), the use of the energy production rates (2.17), (2.18) and (2.19), a hypothesis on the initial chemical composition, and consideration of the changes in this composition resulting from the nuclear reactions, allow us in principle to determine completely the density, temperature and chemical composition at different depths inside the star as well as their time evolution. In practice this is a rather difficult problem, so Eddington attempted to simplify it as soon as it was set by him, in the years 1916–1918. It is interesting to reproduce his reasoning, because of its historical interest and also because it put into evidence the principles of the problem (Box 2.1).

Box 2.1. The simplified stellar model of Eddington

Eddington takes as constant the ratio of the luminosity $L(r)$ at radius r to the mass M_r inside r:

$$L(r)/M_r = L/M. \tag{E2.1}$$

This implies that the ratio of the radiation pressure to the total pressure is constant inside the star: it is taken as $1 - \beta$, β being a constant to be determined. The radiation pressure is:

$$P_{rad} = 4/3c\ \sigma T^4 = 1/3\ aT^4, \tag{E2.2}$$

so that the total pressure is, if the matter is assumed to be not degenerate, instead behaving like a perfect gas:

$$P = 1/3\ aT^4/(1 - \beta) = \Re\rho T/\mu m_H \beta. \tag{E2.3}$$

After some simple manipulations, we get:

$$P = k\rho^{4/3}, \text{ with} \tag{E2.4}$$

$$k = [3\Re^4(1-\beta)/a(\mu m_H)^4\beta^4]^{1/3}. \tag{E2.5}$$

Equation (E2.4) together with the equation for hydrostatic equilibrium (2.6) determine completely the equilibrium condition for the star. Assuming the molecular weight μ is constant β is chosen in order to reproduce the radius R and the mass M of the star. For this, we can see easily from the preceding equations that $M^2/R^4 \propto k(M/R^3)^{2/3}$, implying $k \propto M^{2/3}$. Hence the equation giving β is:

$$1 - \beta = \text{constant} \times M^2\mu^2\beta^4. \tag{E2.6}$$

Note that the radius is eliminated in this equation. If μ stays constant during the evolution, so does β.

Let us examine now the transfer of radiation; we assume no convection. The flux f of the radiation per unit area in any layer of the star is related to the radiation pressure by the relation:

$$f = -c/\kappa\rho\ dP_{rad}/dr, \tag{E2.7}$$

(Continued)

where κ is the absorption coefficient, assumed to be constant inside the star. It will be called later the Rosseland mean, κ_R. The luminosity of the star is, applying the preceding equation slightly below the surface:

$$L = 4\pi R^2 f(R). \tag{E2.8}$$

Writing $P_{rad} = (1 - \beta)P$ and using (2.6), we obtain near the surface of the star:

$$L = 4\pi c G M (1 - \beta)/\kappa. \tag{E2.9}$$

This is a mass–luminosity relation. β is adjusted by comparing this equation with observations, with the result that it depends heavily on the mass, as is indeed clear from Equation (E2.6). Experimentally, we have rather $L \propto M^3$ (see Figure 2.5). Eddington tried to improve his model by introducing a variation of μ with radius, but it loses the attraction of simplicity.

The simplified Eddington's model allows us to obtain an order of magnitude for the temperature T_c and the pressure P_c at the center of the star. This was entirely new at that time, when the origin of stellar energy was completely unknown:

$$T_c \approx 20\,\beta\mu(M/M_\odot)\,(R_\odot/R)\,10^6\ \text{K}, \tag{E2.10}$$

$$P_c \approx 1.2 \times 10^{17}\,(M/M_\odot)^2(R_\odot/R)^4\ \text{dynes cm}^{-2}. \tag{E2.11}$$

Note that the source of energy does not appear explicitly in Eddington's model. An energy source is necessary for the initial assumption (Equation E2.1), but the source could be gravitational contraction.

Actually, the approximations of the simplified Eddington's model are not valid: for example, the Rosseland mean κ_R, which is difficult to calculate, varies enormously with temperature, density and element abundances and does not obey simple laws. Also, convection poses big problems: it is not easy to estimate until what radius the convective bubbles rises before they merge with the surrounding medium. We generally introduce a *mixing length* α, which is treated as a more or less free parameter. Moreover, the kinetic energy of the convective cells is such that they can rise above the radius at which the Schwarzschild or Ledoux criterion for convection predicts that the medium is stable: this is the *convective overshooting*, which is also difficult to estimate. For a long time, it has been considered as a free parameter determined empirically by adjusting the models to the observations, until recent numerical simulations made the convection phenomena clearer. Convection tends to homogeneise the chemical composition while gravity tends to draw the heavier elements toward the centre. The rotation of the star, whose angular velocity depends generally of the radius, produces turbulence and affects the convection, making the problem even more complex. Finally the matter can be degenerate and has then a behaviour very different from that of a perfect gas. These difficulties are such that it is impossible in practice to obtain analytical solutions and that we must resort to numerical calculations in order to solve the problems of the structure and evolution of stars. The first realistic models, due to Henyey, coincide with the appearance of computers around 1960.

In the usual modelling, we assumes an initial chemical composition for the star, but a difficulty arises from the start: the abundance of helium, which plays a major role, is not well known. We prefer to take this abundance as unknown, and to determine it through an adjustment of models to observations. Here is how we proceed in general to model the evolution of a star: we start with an homogeneous chemical composition, where we specify the abundance of helium with respect to hydrogen (mass ratio Y_0/X_0) and the abundances of heavier elements that astronomers group under the name of "metals" (mass ratio Z_0/X_0). The mixing length α is also assumed, and the model is left to evolve. For the Sun, whose age (4.57 billion years) is known thanks to isotopic analysis of meteorites which were born at the same time, the comparison of model to observations can be quite detailed. We compare the calculated radius, luminosity and abundances of metals to observations. If there are discrepancies, we resume the calculation with changes in the initial parameters Y_0/X_0, Z_0/X_0 and α until a good agreement is reached. The most recent values for the chemical composition of the photosphere of the Sun are $X = 0.7393$, $Y = 0.2485$ and $Z = 0.0122$. The initial abundances at the beginning of the evolution of the Sun result from model calculations and are $X_0 = 0.7133$, $Y_0 = 0.2735$ and $Z_0 = 0.0132$.[2] These values are smaller than the observed values in our Galaxy in the vicinity of the Sun. This is not unexpected because the material of the Galaxy is continuously

[2] One could however be surprised that the surface abundance of helium has decreased with respect to the initial one in spite of the fusion of hydrogen. This is due to the sorting of elements by gravity, which drives elements heavier than hydrogen towards the centre.

enriched in helium and heavy elements by stellar nucleosynthesis, as we will see in detail in Chapter 6. Moreover, stellar migration in the Galactic disk is such that the Sun was probably not born at the radius of the galactic disk where it is located at present.

Ideally, modelling of stellar evolution should start with the very young star, before nuclear reactions start. But, for simplification, we assume in general that hydrogen fusion, which begins at the centre of the star where temperature and density are maximal, has already reached equilibrium. The star is then on the *zero age main sequence* (ZAMS).

2.6 Normal and degenerate matter

Before describing in the next chapter the evolution of stars of different masses, it is useful to say a few words on degenerate matter. In effect, the pressure and temperature in the central regions of certain stars, or even in the whole star for white dwarfs and neutron stars, are such that the matter is not in the normal phase of an almost perfect gas. Figure 2.7 schematises the different possible phases of matter.

The grey zone in this figure is where the matter is practically a perfect gas, and where the thermal pressure dominates over the radiation pressure. The transitions between the molecular, atomic and ionised states of hydrogen are indicated. Note that all the representative points of the present Sun at different depths (bold dotted line) lie entirely within this region of the diagram.

If the temperature rises such that the representative point comes above the oblique bold line on the left, the radiation becomes so intense that the radiation pressure dominates over the thermal pressure. If the temperature goes above 5.9×10^9 K, such that $kT > 0.51$ MeV, the mass energy of the electron, there is equilibrium between gamma-ray photons and $e^-{-}e^+$ pairs. This reduces the specific heats and can decrease the adiabatic exponent γ below 4/3, producing instability, for example in certain supernovae. At a still higher temperature (1.5×10^{10} K), so that kT becomes larger than the difference in the mass energy of a neutron and a proton, i.e. 1.29 MeV, the protons are progressively converted into neutrons (neutronisation) when the temperature increases.

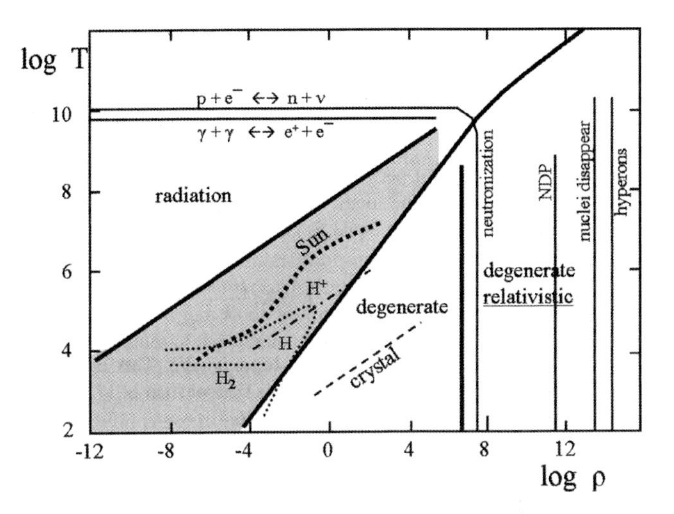

Figure 2.7. The various domains with different equations of state of matter, in the plane $\log T$ vs. $\log \rho$. The bold lines delineate from left to right the zones where the radiation pressure, then the thermal pressure of a perfect gas and finally the non-relativistic degenerate pressure dominate. Explanations in the text. *From Maeder (2009).*

If we now leave the perfect gas domain by increasing the density, the gas becomes degenerate. Quantum mechanics allows only two electrons in an elementary volume of the phase space. If the density increases, the electrons are forced into states of higher energy, so that the pressure does not depend on temperature anymore. For a completely degenerate gas, the pressure is:

$$P_{\mathrm{Fermi}} = K_1(\rho/\mu_e)^{5/3} \text{ dynes cm}^{-2}, \tag{2.20}$$

where μ_e is the molecular weight *per electron*, and $K_1 = 1.00 \times 10^{13}$ in cgs units. This is the case for the matter in white dwarfs.

If the density is still higher, the degeneracy becomes relativistic, and we have:

$$P_{\mathrm{Fermi}} = K_2(\rho/\mu_e)^{4/3} \text{ dynes cm}^{-2}, \tag{2.21}$$

with $K_2 = 1.24 \times 10^{15}$ in cgs units.

When the electrostatic forces between ions dominate the thermal energy, these ions do not experience thermal motions and there is crystallisation. Terrestrial materials are often in this state.

The increase of density in a degenerate medium produces neutronisation of the protons when the Fermi energy of electrons becomes larger than 1.29 MeV. There are more and more neutrons, and their capture by the nuclei produce other nuclei even richer in neutrons. This characterises the neutron stars. What occurs at even larger densities is not of interest for stellar physics.

The degeneracy of matter has deep consequences for the physics and evolution of stars. Let us discuss for example the relation between the mass and radius of white dwarfs where the matter is fully degenerate. The internal pressure can be estimated simply from the equation of hydrostatic equilibrium (2.6): $dP(r)/dr = -GM_r\rho(r)/r^2$. In order to get an order of magnitude, let us take $|dP(r)/dr| \approx P_c/R$, P_c being the pressure in the central regions, and $M_r \approx M/2$, $r = R/2$, $\rho \approx 3\,M/4\pi R^2$, giving:

$$P_c \approx 3/2\pi GM^2/R^4, \tag{2.22}$$

an expression similar to (E2.11), but with a different numerical factor. Writing that P_c is equal to P_{Fermi} in order to insure equilibrium, we find:

$$R/R_\odot \approx 0.012\,(M/M_\odot)^{-1/3}(\mu_e/2)^{-5/3}. \tag{2.23}$$

For a white dwarf, $\mu_e \approx 2$ and we have the apparently paradoxical result that the radius is smaller if the mass is larger: this shows that the properties of a degenerate gas are completely different from those of a perfect gas. For a white dwarf of 1 $M_\odot$, the radius is 0.01 $R_\odot$, approximately equal to the radius of the Earth.

It is obvious that the radius cannot decrease indefinitely, or the mass grow forever. If the radius decreases, the density increases and the degeneracy becomes relativistic. We have then, equating P_c to P_{Fermi} (Equations (2.21) and (2.22))

$$GM^2/R^4 \approx K_2/\mu_e\, M^{4/3}/R^4. \tag{2.24}$$

R^4 cancels, indicating that the mass is constant. This mass is an upper limit, because if it were to grow, the gravitation (the term on the left) would increase faster than the Fermi pressure and the star would collapse. This is the *Chandrasekhar mass limit.* More sophisticated calculations show that this mass is of the order of 1.4 $M_{\odot}$, in good agreement with the highest masses observed for white dwarfs.

By a similar reasoning, we can show that there is an upper mass limit for neutron stars: this is the *Oppenheimer: Volkoff mass limit*, which is of the order of 2 to 3 $M_{\odot}$: its exact value is rather uncertain, due to our imperfect knowledge of the properties of neutronic matter.

2.7 Stellar oscillations

Of course, it is impossible to observe directly the interior of a star. How then could we check the validity of models? We have seen in Section 2.4.1 than neutrinos can give information on nuclear reactions in the core, provided that the transformation between the different species of neutrinos is taken into account. However, this can only be applied to the Sun and, as we will see later, to supernova explosions. Fortunately, observation of stellar oscillations can give very precious information on the structure and even the inner rotation of stars.

2.7.1 Radial pulsations

First of all, certain stars like the cepheids, named from their prototype β Cephei, exhibit radial oscillations, usually designated as pulsations: their radius varies regularly with a period from a few days to a few weeks. These oscillations are generated by thermal instabilities. There are other kinds of periodic variables, in particular the giant stars named Miras (from the name of their prototype omicron Ceti, nicknamed Mira Ceti). Their periods are of several hundred days. Pulsations can occur in the fundamental or harmonic modes. An order of magnitude estimate for the period of the fundamental mode is the crossing time of a pressure wave through the star:

$$p_0 \approx 2R/c_S, \tag{2.25}$$

c_S being the velocity of sound: $c_S = (\gamma P/\rho)^{1/2}$, P being the pressure and $\gamma = C_P/C_V$ the adiabatic exponent. In order to obtain an approximate value of the pressure, we can assume that at any point this pressure balances the weight of the material above. For example, at the centre, g being the gravitational acceleration and the sign $< >$ indicating average values in the star:

$$P \approx <\rho>R<g> \approx <\rho>R(GM/R^2) = <\rho>GM/R, \tag{2.26}$$

which gives for the period:

$$p_0 \approx 2/(\gamma GM/R^3)^{1/2} = 2/[(4\pi/3)\gamma G<\rho>]^{1/2}. \tag{2.27}$$

A more rigorous calculation would yield:

$$p_0 = [3\pi/(3\gamma - 4)G<\rho>]^{1/2}, \tag{2.28}$$

with $<\rho> = M/(4\pi/3)R^3$, the average density of the star. This is still an approximate formula because in general γ is not constant at different radii in the star. The harmonic modes are such that $p_1/p_0 \approx 0.755$, $p_2/p_0 \approx 0.605$ and $p_3/p_0 \approx 0.506$. These ratios differ from 1/2, 1/3 and 1/4 respectively because the velocity of sound between the surface and the first node of vibration differs for the different harmonics since it depends on $\sqrt{T}$. The pulsation period, either fundamental of harmonic, depends on the stellar mass and radius like:

$$p \propto R^{3/2}/M^{1/2}. \tag{2.29}$$

Using the mass–luminosity relation $M \propto L^{1/\alpha}$ ($\alpha \approx 3.3$ for cepheids) and the definition of the effective temperature such that $R^{3/2} \propto (L/T_{eff}^{\ 4})^{3/4}$, we get:

$$p = QL^{(3/4-1/2\alpha)}/T_{eff}^{\ 3}, \tag{2.30}$$

Q being a numerical factor which varies slightly in the Hertzprung–Russell diagram: this is the period–luminosity relation for cepheids.

2.7.2 Solar oscillations

A star is a sphere of gas in hydrostatic equilibrium between pressure and gravity. This sphere behaves like a gigantic resonator, which vibrates in two different ways. On the one hand we find acoustic waves, where the restoring force is pressure (*p waves*): radial pulsations are examples of such waves. They are analogous to the longitudinal seismic waves in the Earth. On the other hand, there are gravity waves (*g waves*), for which the restoring force is the Archimedes force, due to local differences in density. They propagate badly in convective regions, so that inside the Sun, they mainly exist in the deep radiative regions, for which they are unfortunately the only ones which can give information. All these waves are generated naturally in the star either by thermal instabilities or by the turbulent motions of the convective zones. Like the natural or artificial seismic waves for the Earth, the acoustic waves of the Sun give the profile of the sound velocity with depth. The gravity waves, on their side, give the profile of the Archimedes force in the radiative zones.

The resonances of the sphere are such that only certain discrete frequencies can propagate without being rapidly attenuated, in the same way that only discrete resonant sounds can be produced by a vibrating string or plate. The p waves propagate inside the Sun as shown in Figure 2.8, and some examples of resonant vibration modes are presented in Figure 2.9. As the excitation sources have a random distribution, the waves are also distributed randomly and only their frequencies are meaningful. This frequency ranges from one vibration in 3 minutes to one vibration per hour for p waves, with a peak around one vibration in 5 minutes. For the g waves, it is of the order of one vibration per hour. The surface of the Sun oscillates simultaneously in millions of p modes, whose individual amplitude at the surface can reach a few tens of centimetres per second. The frequency of the waves is a signature of the structures they cross. As can be seen in Figure 2.8, the different waves propagate and give in this way information on the refractive index of the different layers, to different depths hence and also on some physical parameters. The rotation of the Sun produces a splitting of the frequency in several components, due to the Doppler–Fizeau effect: the waves coming from one half of the Sun are shifted in one direction, while those from the other half are shifted in the

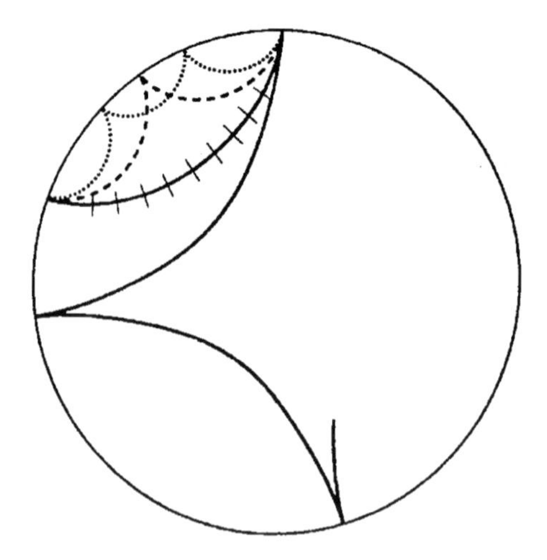

Figure 2.8. Scheme of the propagation of *p* waves inside the Sun. They are created near the surface and penetrate inside; being refracted by the gradient of the refractive index, they come up again to the surface, where they are reflected to the interior, etc. Only the resonant waves, which come back to their origin after a whole number of reflections, are intense. On one of these waves, which are longitudinal, the wave fronts are indicated. We should realise that millions of simultaneous such waves are excited at different points of the surface of the Sun. *From Sylvie Vauclair, with thanks.*

other direction: we can in this way obtain information on the rotation of the different layers of the Sun.

At the present time, only the pressure waves have been well observed and studied in detail either by photometry or by the Doppler–Fizeau effect in the spectral lines of the photosphere. The variation with depth of the velocity of sound is shown in Figure 2.10: we observe a change of slope at a radius of 500 000 km, which corresponds to the limit between the external convective zone and the internal radiative one, and thus gives the value of this radius.

It is much more difficult to observe the gravity waves, because their amplitude at the surface of the Sun should reach at most a few millimetres per second, due to their strong attenuation before arriving at the surface. They have perhaps been marginally detected, but this requires confirmation.

What is known of the rotation of the Sun is represented in Figure 2.11. At the surface, the rotation period varies with latitude from 25.5 days at

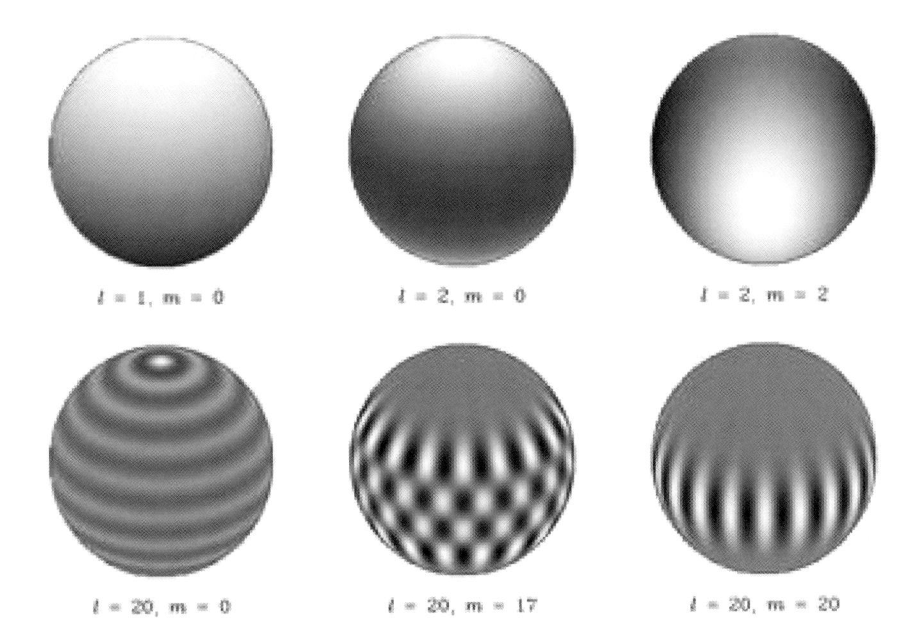

Figure 2.9. A few proper modes of vibration of the Sun. These are *p* modes, defined by two orthogonal integers *l* and *m*, similar to quantum numbers. At a given time, the clear zones expand and the dark ones contract. The reverse happens after half a period. Top left, the Sun has the shape of a pear alternatively elongated to the top or to the bottom; middle top, it oscillates between the shape of a rugby ball and that of a flattened ellipsoid, etc. The actual vibration is the superposition of all these modes which are excited everywhere in the convective zone. They are oriented randomly, and their amplitudes and phases are also random. Only their frequency, which depends on *l* and *m*, has a significance. *From Christensen-Dalsgaard, J. (2002) Reviews of Modern Physics, 74, 1083.*

the equator to 30 days at a latitude of 60° and 34 days in the polar regions. This differential rotation also exists in the convective zone, while the internal radiative zone rotates like a solid body and the nucleus probably rotates faster. These facts are very important for understanding the mixing of matter inside the Sun and the generation of magnetic fields. We will come back to these points in the next chapter.

The solar oscillations cannot only be used to measure the structure and rotation of the Sun, but also to test the ensemble of the parameters of the reactions of hydrogen fusion. For example, the global cross-section of the proton–proton cycle has been determined with a precision of 1%, which is completely out of reach of laboratory measurements.

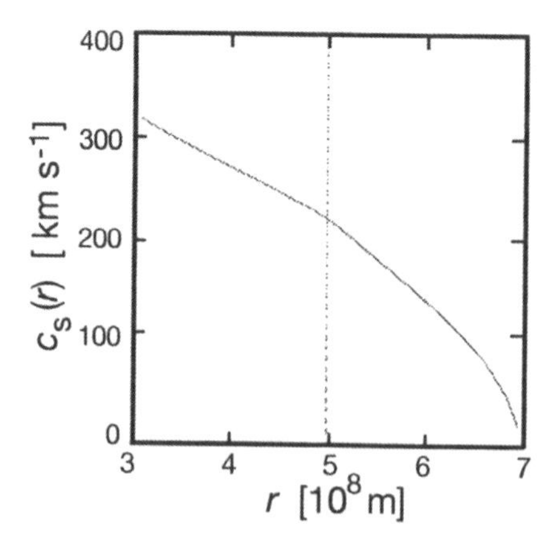

Figure 2.10. Variation of the velocity of sound with depth in the Sun, as derived from a study of the p waves. *From de Boer & Seggewiss (2008).*

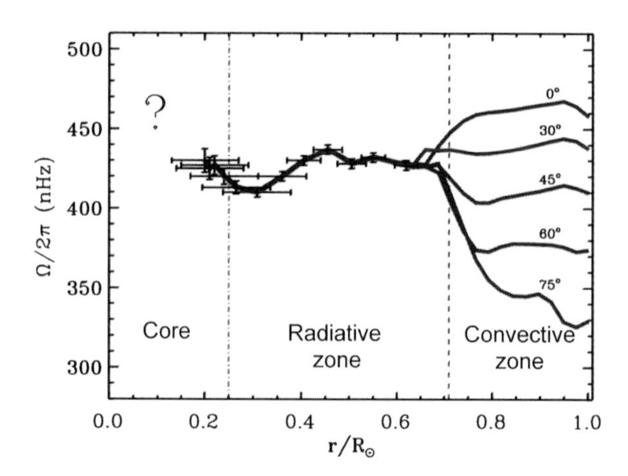

Figure 2.11. The rotation of the Sun. The frequency, in nanohertz, is plotted as a function of radius. 400 nHz corresponds to a period of 28.94 days. In the convective zone, the rotation frequency depends on the latitude, which is indicated for each curve. The rotation of the core is unknown, but is probably faster. *From Turck-Chièze, S. (2009) in Stellar Magnetism, Neiner, C. & Zahn, J.-P., eds., EAS Publication Series 39, 69.*

2.7.3 Stellar oscillations

The study of stellar oscillations, *asteroseismology*, is of course less advanced than that of solar oscillations, *helioseismology*. From the ground, it is only possible to measure global radial velocity variations, which are very weak. However, individual p modes have been identified for the first time in 2002 for α Centauri A, a G2V star slightly hotter than the Sun, and their frequencies are in very good agreement with theoretical predictions. The situation is rapidly evolving thanks to the CoRoT, MOST and KEPLER satellites, which measure periodical light variations of many stars to amplitudes as small as 10^{-6} in relative value. Many results have been obtained, but we cannot discuss them here, especially because the situation is evolving very rapidly.

3

The Evolution of Single Stars

3.1 The evolution of the Sun

The evolution of a solar-mass star exhibits all the stages we will find for stars of other masses (Figure 3.1). Starting from the ZAMS where it stabilises after the pre-main sequence phase, the star enters the longest stage of its life, the main-sequence stage, during which hydrogen is converted into helium in its core through the proton–proton mechanism. For the Sun, this phase will last 11 billion years; at its present age of 4.57 billion years, it has spent somewhat less than half its life on the main sequence. The present structure of the Sun is shown in Figure 3.2.

During the main-sequence phase, the luminosity of the Sun will double while its radius will increase by about 50%, whilst its effective temperature will not change much. Of course, this is not observable for the Sun, but we know enough 1 $M_\odot$ stars at different stages of evolution to check this prevision of the models. The representative point for the Sun in the Hertzprung–Russell (HR) diagram will not move much during its evolution on the main sequence. Note that during hydrogen fusion, the number of particles decreases: the medium being fully ionised, the result of the p–p chain is:

$$4\,H^+ + 4e^- \rightarrow 1\,He^{++} + 2e^- + 4\gamma + 2\nu_e + \text{energy}.$$

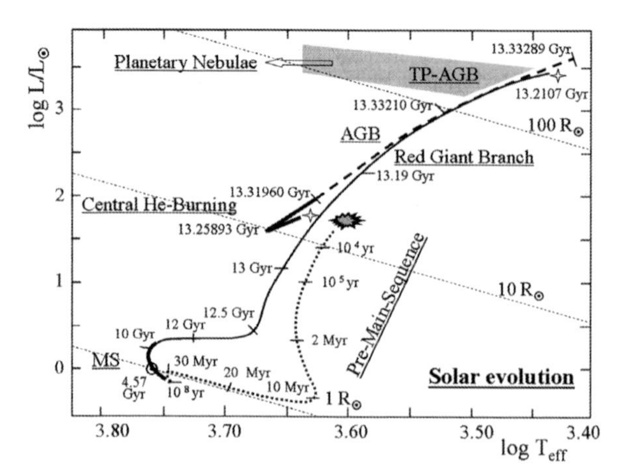

Figure 3.1. The evolution of the Sun. The evolution track of a solar-mass star is shown in the HR diagram from its formation to its death as a planetary nebula. Time is indicated at different steps along the track. The two 4-branch stars correspond to the helium flash and to the subsequent rapid rearrangement of the structure of the star. *From Maeder (2009), data from Corinne Charbonnel.*

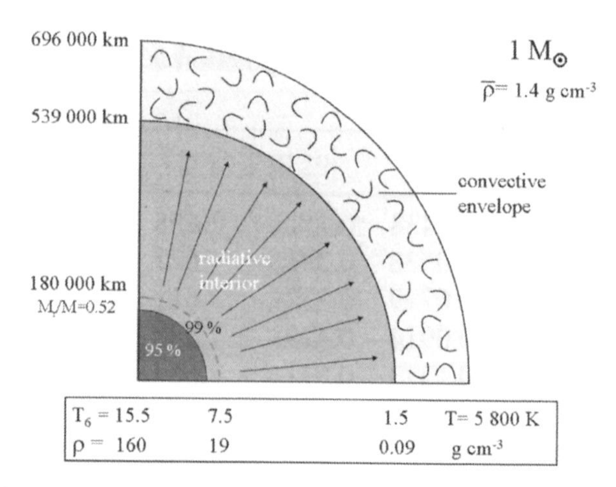

Figure 3.2. The present structure of the Sun. The physical parameters are indicated in the figure. The temperature T_6 is in million K. *From Maeder (2009).*

From 8 particles, we are left with only 3, so that the mean molecular weight increases. This produces a gradual contraction and heating of the core, hence a progressive acceleration of the fusion, which however does not become explosive. More and more external zones heat up. The mass

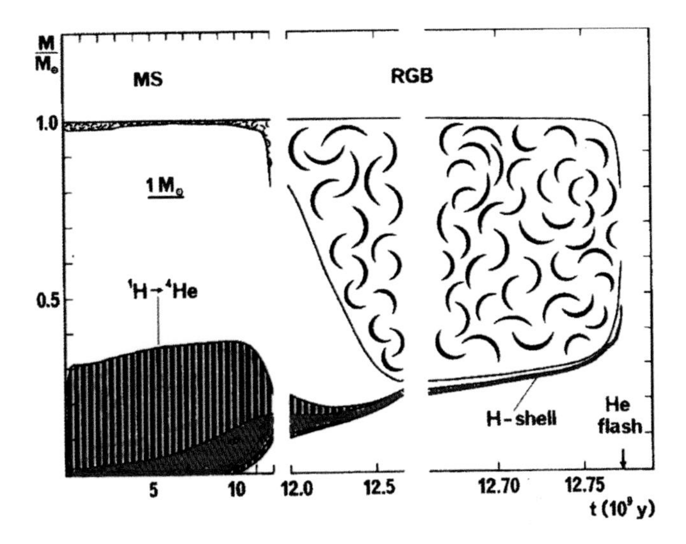

Figure 3.3. Evolution of the structure of the Sun on the main sequence and on the giant branch. The ordinates give the mass fraction from the centre (0) to the surface (1.0): compare with Figure 3.2 where the radius is plotted linearly. Time is in abscissas (note the different scales). The cloudy regions are convective, otherwise the energy transport is radiative. The region of energy production by hydrogen fusion is marked by lines, with a maximum in red (more than 10 erg g^{-1} s^{-1}). *From Maeder, A. & Meynet, G. (1989) Astronomy & Astrophysics 210, 155, with permission of ESO.*

of the fusion zone increases slowly during the stay on the main sequence. Throughout this stage, energy is transported outwards by radiation deep inside the Sun, and by convection in the external regions (Figure 3.3). This convective region, with its magnetic field, is responsible for the solar activity.

When hydrogen is exhausted at the centre, combustion goes on in a layer surrounding the helium nucleus, which is now inert. By convention, we say that the star is leaving the main sequence. There is no discontinuity in the evolution, but it becomes faster. The combustion layer displaces to the outside, while the helium nucleus contracts and its matter becomes degenerate. The energy production increases progressively, yielding an expansion of the envelope, and the star becomes a red giant: red, because the increase of the radiated energy is not sufficient, given the enormous increase of the radius, to maintain a high external temperature. In the Hertzprung–Russell diagram, the figurative point of the star follows the *red giant branch* (RGB).

The expansion of the envelope produces cooling of its external regions. Also, the heat from inside takes time to reach them. Some ions thus recombine with free electrons, producing a strong increase of opacity which accelerates the expansion. The temperature gradient becomes strong at the level of recombination, triggering convection. Now the transport of energy from the interior is favoured, accelerating expansion. The convection carries to the surface internal matter whose chemical composition has been modified by nuclear reactions: this is called the *dredge up*.

13.2 billion years after the ZAMS, the *helium flash* occurs. The temperature at the centre of the core is now of the order of 10^8 K, so that helium starts to transform itself into carbon by the 3α reaction (see section 4.3 of the preceding chapter). This reaction is explosive because the matter is now strongly degenerate, the density reaching 10^6 g cm^{-3}: an increase in temperature does not yield an increase in pressure like for a perfect gas, so that nothing controls for the moment the reaction. An enormous quantity of energy is liberated in a few seconds deep inside the core. Nothing can be seen at the surface, because of the time necessary to transport the energy outside. Moreover, a large part of this energy is in the form of neutrinos which escape freely from the star. In the core, the temperature has increased so much that the degeneracy of matter is raised: the core can expand and the combustion of helium becomes quiet. The structure of the star readjusts itself, with the consequence that the production of energy by fusion of hydrogen decreases, and a process inverse of that producing red giants occurs: the luminosity and radius decrease rapidly until a new stable configuration sets up.

The star has now a layer structure: around the core which is again inert, and which consists essentially of carbon, there is a thin layer where helium burns (Figure 3.4). Then we find a thick layer of helium, with another thin layer above where hydrogen burns. Above is the convective envelope. The extra energy produced by the combustion of helium is such that in the HR diagram the star follows a track close to that which it followed previously as a red giant: this track is called the *asymptotic giant branch* (AGB). The beginning of this phase is the E-AGB one (E for *early*).

The combustion of helium in a thin layer is unstable, so that thermal pulses occur at the end of the AGB phase: this is the TP-AGB phase (TP

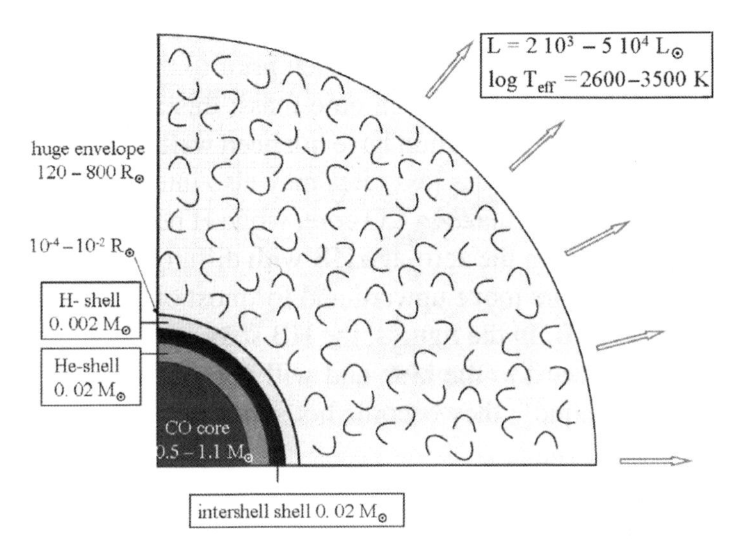

Figure 3.4. Structure of a star with an initial mass 2.5 $M_\odot$ at the beginning of the asymptotic giant branch. That of a 1 $M_\odot$ star is qualitatively similar. The envelope is fully convective in this case. *From Maeder (2009).*

for *thermal pulses*). These pulses generate a strong mass loss, a mass loss that had already started during the ascent of the giant branch. In the expanding envelope generated by these pulses, the temperature decreases so that first molecules can form, then solid particles. The pulses also produce dredge-ups of the internal matter, whose chemical composition has been strongly modified by nuclear reactions. Figure 3.5 shows the result of numerical simulations for a 2.5 $M_\odot$ star; the result is qualitatively the same for other masses.

We can see on Figure 3.5 that the radius and luminosity of the star increase strongly during the TP-AGB phase, while its effective temperature decreases. The star becomes a red supergiant. It exhibit important radial pulsations that further increase the mass loss. The expanding envelope becomes fully opaque because its decreasing temperature allows the formation of molecules and of dust grains, and it is very difficult to see what occurs under the surface. We will come back to this point in the next chapter.

Eventually, after having expelled almost half its matter, the star is reduced after the last pulse to its core of carbon and oxygen, the latter

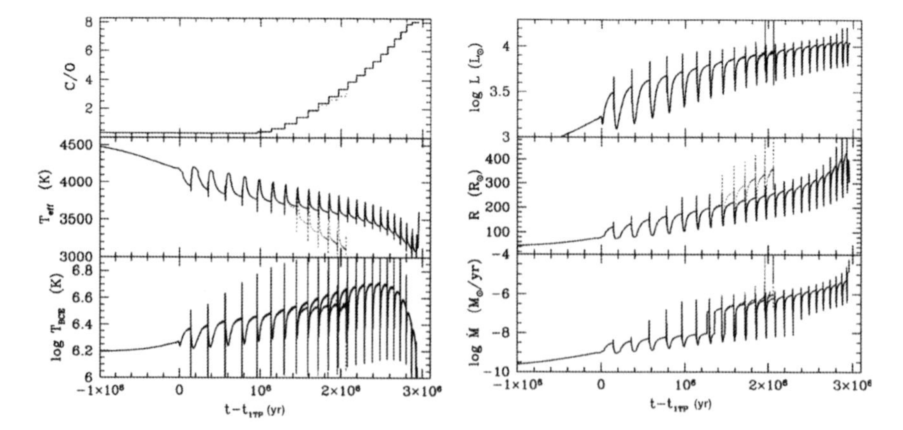

Figure 3.5. The evolution of a 2.5 $M_\odot$ star during the TP-AGB phase, where thermal pulses occur. Left and from top to bottom: the evolution of the abundance ratio of carbon to oxygen at the surface of the star, then of effective temperature and finally of the temperature at the bottom of the convective envelope. Right, from top to bottom: the evolution of luminosity, then of radius, and finally of mass loss. Time is counted from the beginning of the pulses. *From Decressin, T. (2007) reproduced by Maeder (2009).*

being formed by the reaction $^{12}C + {}^{4}He \rightarrow {}^{16}O$. This core, which is almost bare and extremely hot, has a mass of about 0.55 $M_\odot$ for an initial mass of the star of 1 $M_\odot$. It is surrounded by an expanding envelope whose chemical composition is enriched in carbon. The matter of the inner parts of this envelope is strongly ionised by the ultraviolet radiation of the central star and becomes easily visible: this is a *planetary nebula*. The external parts of this nebula are made of neutral gas, partly molecular, and of dust condensed from this gas. Then the central star, which is deprived of energy sources, condenses rapidly to become a white dwarf, while the envelope is dispersed into the interstellar medium.

3.2 The evolution of stars of small or intermediate mass

Figure 3.6 shows the evolutionary tracks calculated for stars of different masses in the "theoretical" HR diagram.

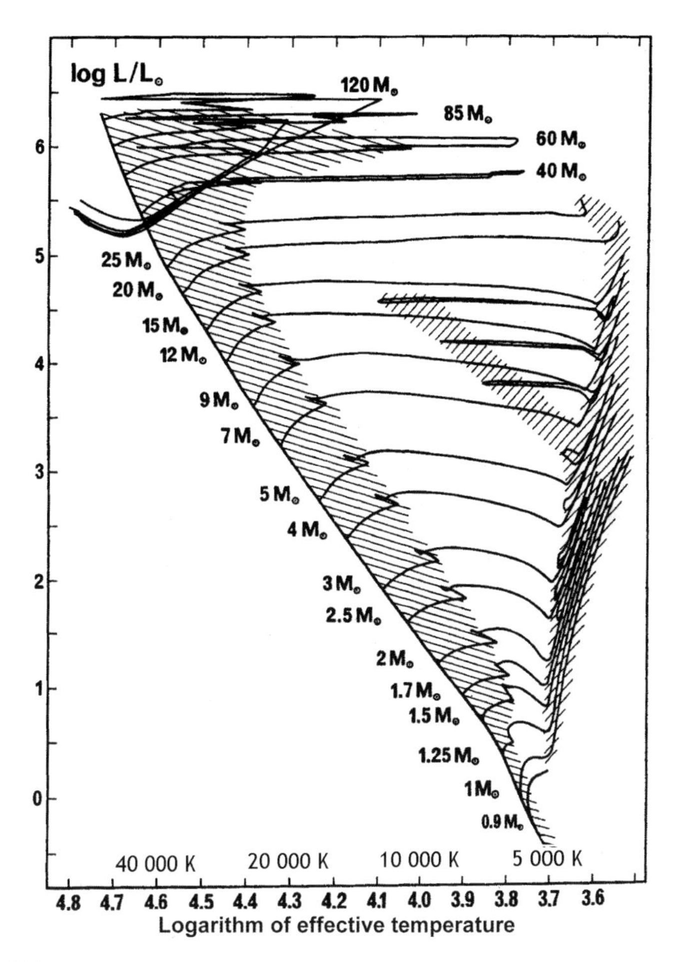

Figure 3.6. Evolutionary tracks of stars of different masses in the "theoretical" HR diagram. *From Schaller, G. et al. (1992) Astronomy & Astrophysics Supplement Series 96, 269, with permission of ESO.*

The stars whose mass is smaller than that of the Sun have a similar, but slower evolution. The lifetime of a main sequence star of 0.8 $M_\odot$ approaches the age of the Galaxy, i.e. about 13 billion years, so no red giant has yet been produced by these stars.

If the mass of the star is larger than 1.15 $M_\odot$, its central temperature is high enough for hydrogen fusion to occur through the CNO cycles, which occurs through the p–p process for smaller masses. The temperature gradient in the core is then very steep, triggering convection. This convection mixes the matter of the core and carries towards its central parts material rich in hydrogen, while the products of the fusion are carried to the envelope. Hydrogen fusion could therefore last longer than for the smaller stars, at least in principle. However, when the mass of the core reaches about 0.5 $M_\odot$ of helium, there is not enough hydrogen left at the centre to maintain the fusion. Then the central temperature decreases and the core contracts, producing a new emission of energy of gravitational origin. During this episode, the effective temperature of the star as well as its luminosity rises slightly, producing a turn in its evolutionary track (see Figure 3.6). The energy from the contraction heats sufficiently the edge of the core so that the fusion of hydrogen resumes in this region; the star is becoming a red giant. The rest of the evolution is similar to that of a 1 $M_\odot$ star. As an example, Figure 3.7 shows the evolution of a 3 $M_\odot$ star before the TP-AGB phase. For the latter, see Figure 3.5 which corresponds to a slightly smaller mass.

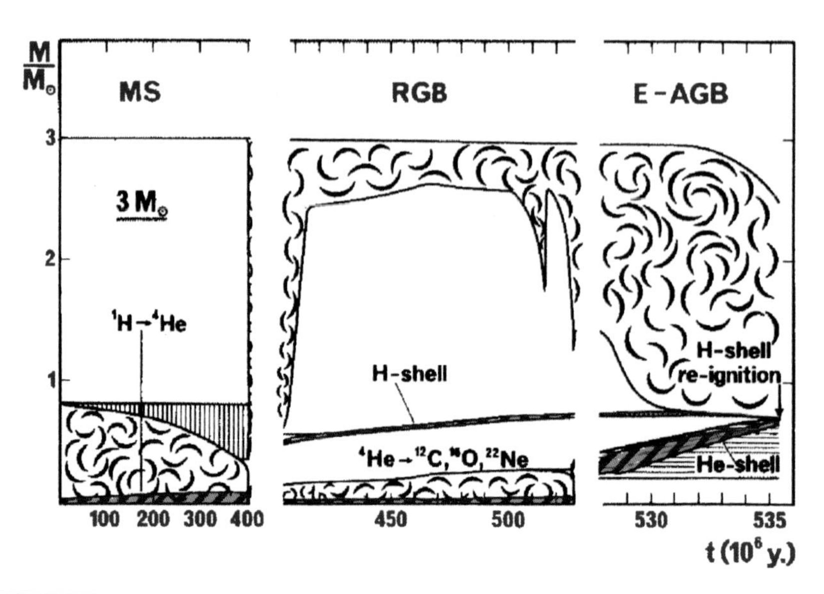

Figure 3.7. Evolution of a 3 $M_\odot$ star. The presentation is the same as for Figure 3.3. The zones of production of energy are in red. The beginning of the evolution on the asymptotic giant branch is indicated (E-AGB). *From Maeder, A. & Meynet, G. (1989) Astronomy & Astrophysics 210, 155, with permission of ESO.*

For masses between about 6 and 15 $M_{\odot}$, a new phenomenon appears: the representative point of the star in the HR diagram makes a loop when helium burns in a layer, after the helium flash. During a part of the stay on this loop, in which the envelope becomes temporarily radiative, the star is unstable and experiences radial pulsations whose period is linked to its luminosity and to its mass (see above Section 2.7.1).

The initial chemical composition of the star affects rather deeply all these phenomena, due to its considerable impact on the opacity of stellar matter. In particular, the evolution of stars with a very low *metallicity* (abundance in heavy elements) is different from that of metal-rich stars when they become giants: after the helium flash, when they burn helium in a layer around the core, they occupy for some time a part of the HR diagram called the *horizontal branch* (Figure 3.8). The physics of this phase is not yet fully understood.

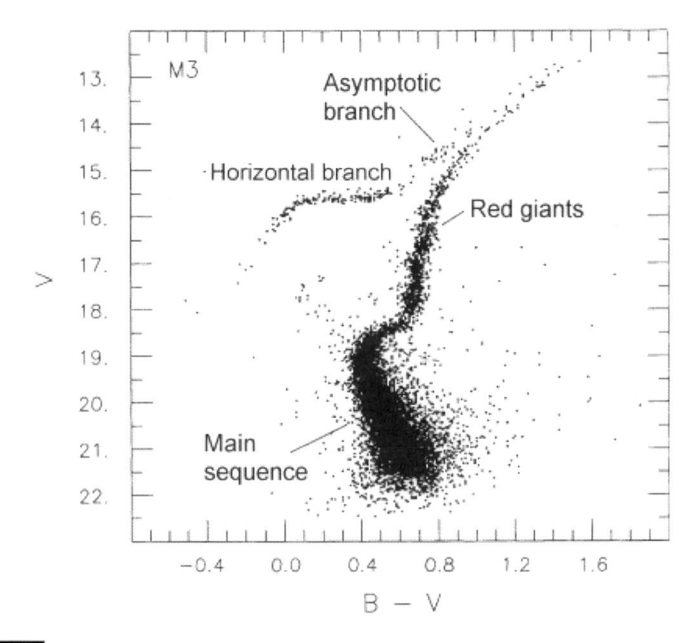

Figure 3.8. The HR diagram of the globular cluster M 3, where the different phases of stellar evolution are indicated. All the stars have the same age, about 13 billion years. Those who have left the mains sequence have all nearly the same mass, about 0.9 $M_{\odot}$. We recognise the giant branch, the AGB and the horizontal branch, characteristic of low-metallicity stars. Compare to Figure 2.3. *Data from Buonanno, R. et al. (1987), in ESO workshop on stellar evolution and dynamics in the outer halo of the Galaxy, p. 331.*

3.3 The evolution of high-mass stars

The evolution of stars with masses larger than 15 $M_\odot$ is not very different in its principles from that of lower-mass stars. When hydrogen is entirely burned in the core, the star leaves the main sequence and becomes a very big and luminous red giant, a *red supergiant*. At this stage, ^{4}He is burning in the core to give ^{12}C by the 3α reaction, with carbon itself partially transformed into ^{16}O, ^{20}Ne and ^{24}Mg through capture of ^{4}He. Then, when the central temperature is over 5×10^8 K, two ^{12}C atoms can fuse to form different nuclei, in particular ^{20}Ne. Moreover, hydrogen burns in a layer above the nucleus by the CNO cycles, which produce a large amount of ^{14}N because one of the reactions of these cycles, $^{14}\text{N} + {}^1\text{H} \rightarrow {}^{15}\text{O} + \gamma$, is very slow. Finally, the star explodes as a supernova. The left part of Figure 3.9

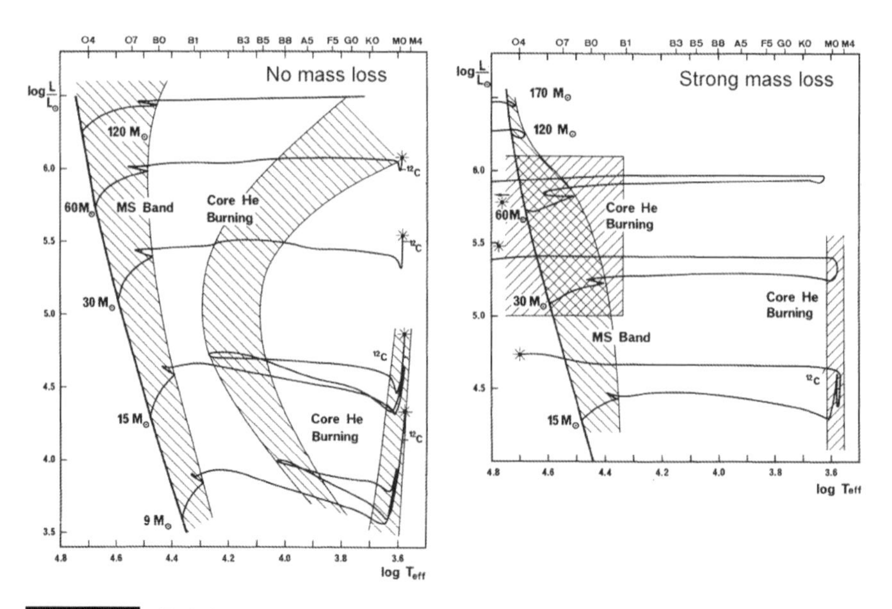

Figure 3.9. Evolutionary tracks in the HR diagram for high-mass stars. Left, without mass loss and right, with a large mass loss. These tracks end by the explosion of the star as a supernova. The regions where hydrogen fusion occurs is in the core (main sequence) and those where helium burns are indicated by lines. The small transverse lines with the mention "^{12}C" corresponds to the beginning of combustion of pure carbon, $^{12}\text{C} + {}^{12}\text{C}$. *From Maeder, A. (1981) Astronomy & Astrophysics 102, 401, with permission of ESO.*

shows the evolutionary tracks of stars of different masses, assuming that there is no mass loss during their evolution. However, the evolution is deeply affected if there is mass loss, which starts during the main sequence phase.

What is the origin of the stellar winds which produce the mass loss in massive stars? A photon absorbed in the resonance line of an atom of the stellar atmosphere gives to this atom a momentum in the direction of its arrival, equal to $h\nu/c$ where ν is the frequency, h the Planck constant and c the velocity of light. It is soon reemitted at the same frequency, the atom losing the momentum $h\nu/c$, but now in a random direction. If the radiation field is not isotropic, as near the surface of a star, this results in a net outward acceleration of the atom. For non-resonance lines, the physics is more complicated because we have to take into account all the energy levels to which the atom falls after it is excited by the incoming photon.

Similarly, an anisotropic continuum produces a net acceleration of atoms and of free electrons. This can occur during the ionisation of an atom by a continuum photon or, for the electrons, via Thomson scattering: this scattering can be considered as the temporary absorption of a photon which is immediately reemitted in a random direction. The electrons accelerated in this way drag by collisions the ions in the vicinity, then these ions themselves drag the neutral atoms by collisions. In this way, a free electron carries with it a mass m_f of the order of the mass of a proton m_p for a fully ionised medium, but a mass m_f 2 to 2.5 times larger in a real, partly ionised hot atmosphere (m_f can be even larger if the atmosphere is poor in hydrogen). Thus, ignoring for a moment the other mechanisms, the matter experiences an acceleration:

$$a = \sigma_e/m_f \, F/c, \tag{3.1}$$

where $\sigma_e = 6.65 \times 10^{-25}$ cm^2 is the Thomson scattering cross-section and F the flux per unit area of the radiation emerging from the surface of the star, integrated over frequency. This flux is simply:

$$F = L/4\pi R^2. \tag{3.2}$$

Acceleration by continuum radiation is very important for massive stars, whose atmosphere is partly ionised given their strong flux in the far ultraviolet. It is of interest to compare the acceleration a to that of gravity at the surface of the star, $g = GM/R^2$:

$$a/g = \sigma_e / 4\pi c G m_f \, L/M \approx 2 \times 10^{-5} \, L/M \, , \tag{3.3}$$

taking $m_f = 2.5\ m_p$; L and M are here in solar units. The atmosphere is unstable if $a > g$, hence for $L > L_{Eddington} = 5 \times 10^4\ M/M_{\odot}$. This Eddington luminosity $L_{Eddington}$ makes sense only if Thomson scattering is the principal mechanism. However, the other mechanisms can also be important, reducing the critical luminosity. We notice, comparing the hot, massive stars in our Galaxy with those of the Magellanic Clouds, which are less rich in heavy elements, that the mass loss of these stars increases a lot with metallicity, showing the importance of acceleration in spectral lines. In the Galaxy, mass loss is already quite large for stars with an initial mass larger than about 15 solar masses. The evolutionary tracks of Figure 3.6 take this into account.

The most recent stellar models include both the interior and atmosphere of stars and can reproduce not only their evolution but also their spectrum at the different steps. This spectrum is characterised by the particular aspect of the lines in the ultraviolet, of which Figure 3.10 gives examples. In rare cases, this shape can be seen for visible lines, for example in the star P Cygni which gave its name to this particular type of line profile.

The right part of Figure 3.9 shows how the evolutionary tracks are modified by mass loss. The stars can partly or entirely miss the red supergiant phase, to move rapidly towards the main sequence. We will see what occurs when we discuss Figure 3.11, which shows the evolution of the structure of a star with an initial mass of 60 $M_{\odot}$.

The main point of evolution with mass loss is that the zones of mixing by convection, and even the combustion zones themselves, can be very close to the surface. For example, we see in Figure 3.11 that the nitrogen produced during the CNO cycle enriches the surface already at the end of the main sequence phase, and even more during the phase of central combustion of helium because the region where CNO occurs is close to the

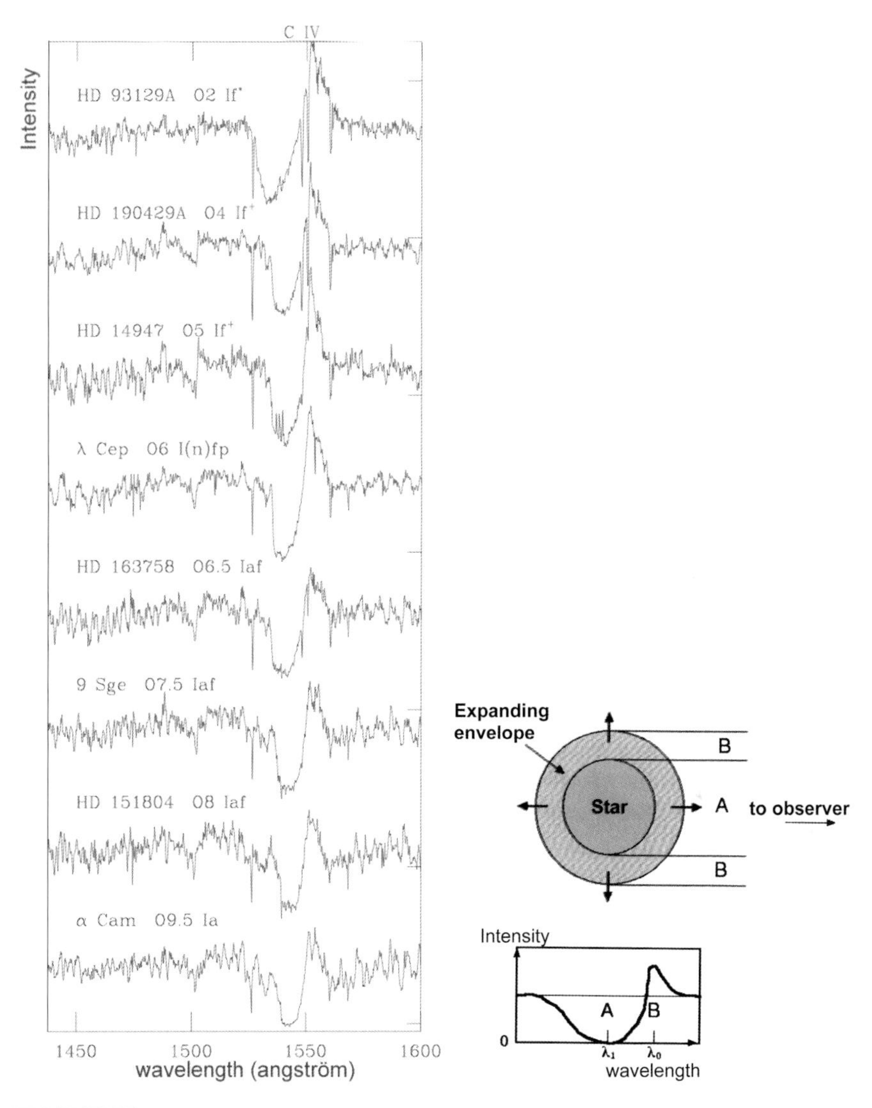

Figure 3.10. Left, portions of ultraviolet spectra of very hot stars containing the resonance line of 3-times ionised carbon (CIV) at 155 nm. The particular shape of this line (the P Cygni profile) is explained on the right. In front of the star (A) its atmosphere absorbs the continuum radiation in the line, which appears in absorption centred at the wavelength λ_1, shifted toward shorter wavelengths since the absorbing gas comes towards the observer. On the edges (B) we see an emission line centred at the wavelength λ_0, not shifted in wavelength. The global spectrum is the superposition of these two spectra. The expansion velocity is $\nu \approx c(\lambda_0 - \lambda_1)/\lambda_0$. *Spectra obtained with the IUE satellite by Walborn, N.R., & Nichols-Bohlin, J. (1987) Publications of the Astronomical Society of the Pacific 99, 40, with permission of the AAS.*

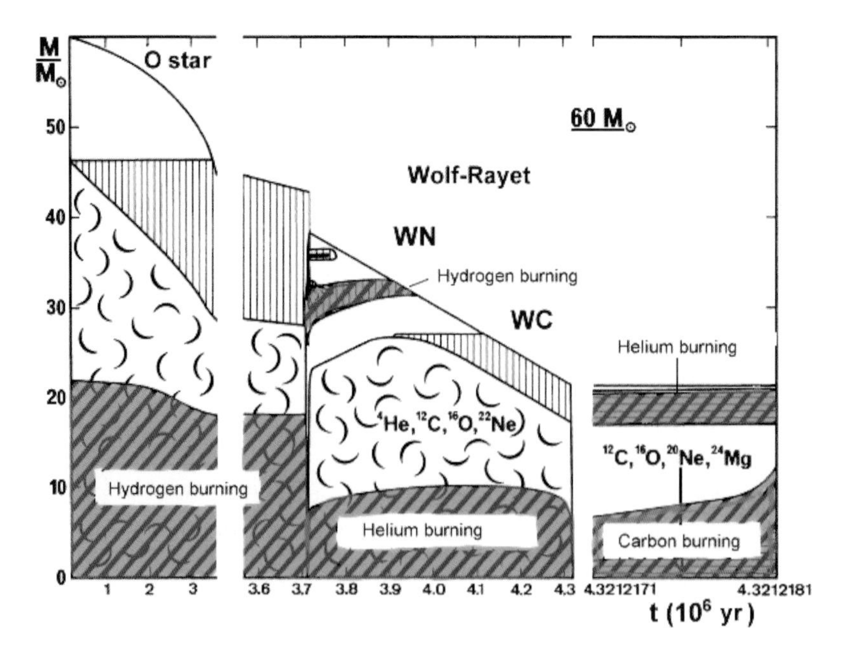

Figure 3.11. Evolution of the structure of a 60 $M_{\odot}$ star with strong mass loss. The star has lost 2/3 of its mass at the end of the evolution. The zones of fusion are shown in red with oblique lines, while those with fine vertical lines correspond to regions where mixing occurs due to convection, and which are this progressively enriched in elements synthetised in deeper regions. After a short red supergiant phase at the age of 3.7 million years, the star becomes a Wolf–Rayet star: first a WN thanks to the nitrogen produced by the CNO cycle in the layer where hydrogen burns, this a WC when the carbon produced in the interior comes to the surface in abundance. The star explodes after a short phase of carbon burning in the central parts: then its mass is only about 20 $M_{\odot}$. *From Maeder, A. & Meynet, G. (1987) Astronomy & Astrophysics 182, 243, with the permission of ESO.*

surface. Later, the carbon and even oxygen produced in the interior of the star come to the surface, and their abundances are considerably higher than that of nitrogen. During all these phases, there is less and less hydrogen at the surface, which is considerably enriched in helium.

What happens to the atmosphere expelled by mass loss? When the star is on the main sequence, the stellar wind, where hydrogen is fully ionised by the radiation of the star, forms a big bubble expanding into the interstellar medium. When the combustion zone of hydrogen reaches the surface of the star, the stellar wind does not contain hydrogen anymore, but rather helium and nitrogen, then carbon and even oxygen for the most massive stars. This hot gas (several ten thousand K) is dense and opaque, so that we cannot see the surface of the star. What we see is a continuum and a few broad lines *in emission*. We have seen in Figure 3.10 that during the main sequence phase the atmosphere outside the edges of the stellar disk produces emission lines, but that the same lines are in absorption in front of the disk, shifted by the Doppler–Fizeau effect. Now, the atmosphere is opaque for the radiation of the disk and the lines are in pure emission, very broadened due to the expansion, whose velocity is of the order of 3 000 km/s: the star has become a Wolf–Rayet. At the beginning, the atmosphere contains mainly helium and nitrogen, whose lines from different degrees of ionisation dominate the spectrum: the star is a WN. Then this wind dissipates and is progressively replaced by a wind rich in carbon instead of nitrogen, and the star becomes a WC. For higher initial masses of stars, oxygen is added to carbon and the star is designated as a WO.

To summarise, our 60 $M_\odot$ star evolves according to the following sequence:

$$O \rightarrow B \rightarrow \text{F–G supergiant} \rightarrow WN \rightarrow WC \rightarrow \text{supernova.}$$

As we have noticed, the effects of metallicity on this evolution are very important. As a result, in galaxies with low metallicities like the Small Magellanic Cloud, the mass loss is small so that the ratio of the number of red supergiants to that of Wolf–Rayet stars is high; conversely, this ratio is small in metal-rich galaxies like ours, because mass loss is such that most massive stars cannot become red supergiants.

Note that practically all massive stars are variable, especially when they become A to G supergiants, hotter than the red supergiants: these are the *long-period variables*.

3.4 Complications: rotation and magnetic field

In the preceding text, we ignored the rotation of stars. However, all stars rotate, because they preserved a part of the angular momentum of their parent cloud. We saw in Section 2.7 that the Sun rotates with a period of 25.5 days at the equator, which increases to 34 days in the polar regions: this is a differential rotation which occurs throughout the convective zone, while the inner radiative zone rotates like a solid body with a period of 27 days, and the core probably has a somewhat faster rotation. Young stars rotate more rapidly than the old ones: for solar-type stars, the rotation period varies with age t as $t^{-0.5}$. Stars with a mass smaller than about 1.4 $M_{\odot}$, which possess a convective envelope, rotate much more slowly than higher-mass stars. There must be some braking of the rotation during the life of stars: we will see that it is due to the magnetic field.

Rotation has considerable consequences on the properties and evolution of stars, to the extent that a full book is devoted to this topic (Maeder, 2009). Observationally, the Doppler–Fizeau effect produces a broadening of the spectral lines, because we observe the superposition of lines from both the receding and approaching regions. A rotating star has the shape of a flattened ellipsoid, so that the temperature derived from the spectral energy distribution depends on its orientation with respect to the line of sight: the polar regions are hotter than the equatorial ones. Let us consider a given solid angle with its summit at the centre of the star. The radiation flux F in this solid angle reaches the pole at a smaller radius than for the equator, and therefore delineates a smaller area of the surface, which is thus raised to a higher temperature and is bluer than the equatorial regions. Roughly, F is proportional to the acceleration g at the surface (*theorem of von Zeipel*, who first noticed this phenomenon in 1924).

Rotation also affects the internal structure. It produces in the radiative zones a slow circulation of matter, which carries towards the exterior chemical elements produced in the interior by nuclear reactions. Conversely, more hydrogen reaches the core, so that the lifetime of the star on the main sequence is increased. Angular motion is transported by the mentioned circulation, as well as by the gravity waves emitted at the limit of the convective zones. We have good reasons to believe that these waves are responsible for the uniform rotation of the radiative interior of the Sun.

These phenomena, and in particular the internal mixing due to rotation, have only recently been considered when building stellar models. It changes the evolutionary tracks in the HR diagram appreciably. Still more important, it changes the internal profile of the chemical composition in late phases of the evolution, so that the composition of the ejected matter is also altered.

Differential rotation like that of the Sun produces a "horizontal" turbulence in the radiative zones, modifying further the internal mixing and triggering various instabilities. But the most remarkable effect is the generation of a magnetic field in the convective zones. Indeed, stars, whose matter conducts electricity, possess in general a magnetic field. It is measured via the Zeeman effect produced on spectral lines.[1] When stars are formed, they have a weak magnetic field coming from the interstellar ones. The differential rotation twists the lines of force of this field, which produces an amplification, compensating the ohmic losses and increasing the global field: this is the *dynamo effect*. This effect is also responsible for the magnetic field of the Earth, whose internal parts are also in differential rotation.

The dynamo effect allows us to understand the time evolution of the magnetic field of the Sun during its 11-year activity period. At the minimum of activity, the magnetic field is weak (about 1 gauss = 10^{-4} tesla), dipolar like that of a magnet and directed along the rotation axis (such a field is called *poloidal*). The differential rotation twists the lines of force. After a few turns, this twist generates progressively by the dynamo effect a magnetic field of several hundred gauss whose lines of force are parallel to the equator, with opposed directions in the two hemispheres: this is a *toroidal* field. This field produces magnetic loops which emerge at the surface, generating sunspots grouped in magnetically bipolar activity regions, where its intensity reaches several thousand gauss. These loops, which are initially parallel to the equator, partly twist in the north–south direction, due to Coriolis acceleration, creating a component along the meridians. These components will merge to form a poloidal field with a direction inverse to that of the initial field, and the next cycle of activity starts with an inverted field. In certain particularly active regions, the field dissipates via a reconnection of the lines of force, producing eruptions. At the beginning of the activity cycle, the loops emerge at high latitudes, then progressively appear at lower latitudes, for reasons which are not yet well understood.

---

[1] Recently, spectropolarimeters have been built which allow us to map the magnetic field at the surface of stars.

Due to the existence of its convective zone, the Sun loses a very small amount of matter (about 10^{-14} $M_\odot$ per year), as a wind whose existence was first suspected from observations of the tails of comets. This wind brings with it some angular momentum, but this loss would be negligible in the absence of magnetic field, which acts as a lever. It is the magnetic field which is responsible for the time decrease of the rotation of solar-type stars. Massive stars do not have a convective zone and thus also little or no magnetic field in general: this explains why they are fast rotators in general. However, a magnetic field is observed in about 5% of them: it is of fossil origin and can reach several thousand gauss. It slows the rotation of these magnetic stars, but now the stellar wind necessary for carrying away the angular momentum is generated by the radiation pressure produced in their atmospheres by their intense radiation.

When the rotation is very fast, the centrifugal force at the equator can generate a strong mass loss and form a rotating disk around the star. This disk reminds us of the protoplanetary disks that we encountered in Chapter 1, but its origin is completely different. The gas in this disk produces emission lines which are superimposed to the usual absorption lines of the star. These stars are Be stars, not to be confused with the very young Be ones.

4

The Death of Stars

While the evolution of stars is rather similar for stars of all masses, there are very important differences at the end of their lives. Below a critical mass of about 8 $M_\odot$, the stars that leave the main sequence become red giants; then, after the helium flash, they climb the asymptotic giant branch. At the end of this stage where they lose most of their mass, they are reduced to a compact object surrounded by the remnants of their envelope. The compact object is relatively inert and becomes a white dwarf which survives forever. Conversely, stars more massive than 8 $M_\odot$ die in a spectacular way: their evolution accelerates dramatically, often by-passing some of the preceding stages, to end by one of the most energetic events of the Universe, a supernova explosion. As a consequence, we have to study separately the stars below or above the critical mass. We will call them respectively the small and the big stars.

4.1 The death of the small stars

4.1.1 After the asymptotic giant branch

In the preceding chapter we left these stars at the end of a series of thermal pulses which characterise the TP-AGB phase. Then, they are surrounded by a thick expanding envelope. They are affected by strong pulsations which increase further the mass loss, which is over 10^{-5} $M_\odot$ per year for about a thousand years. The last thermal pulse produces a quasi-explosive combustion of helium into carbon and oxygen, expelling the envelope which is now detached from the core. Only a few layers still produce energy in the core, but gravity dominates so that it contracts and heats up rapidly: the star goes in a few hundred years from to red supergiant to yellow giant to blue dwarf, while its luminosity stays approximately constant. This post-AGB evolution can only be observed in the infrared, because the star is now hidden by the thick envelope it expelled, whose external parts are sufficiently cold for formation of molecules and grains which make it opaque to visible light.

During the AGB stage, active nucleosynthesis took place in the different combustion layers of the star, whose products can be found in the ejected envelope. Box 4.1 gives details about this nucleosynthesis. The appearance of one or another product in the spectrum of the star is the basis for the following classification:

- most AGB stars are M giants, characterised by a ratio $^{16}O/^{12}C > 1$ and by an excess of ^{14}N produced in the combustion zones of hydrogen through the CNO cycles;
- the rarer S stars are characterised by an excess of ^{14}N, ^{12}C and of some elements produced by neutron capture (see Box 4.1);
- the carbon stars are extreme S stars where the carbon abundance is larger than that of oxygen: $^{12}C/^{16}O > 1$;
- the J stars are carbon stars showing an excess of the ^{13}C isotope.

Box 4.1. Nucleosynthesis in giants of the asymptotic branch

This nucleosynthesis occurs:

- in the thin layer of hydrogen combustion at the end of the E-AGB phase (see Figure 3.7). The temperature is then higher than in the central parts of main-sequence stars, so that combustion is through the CNO cycles; but because the layer is thin the nuclear reactions do not have time to reach equilibrium before some products manage to escape. As a result, there is an important production of not only ^{4}He, but also ^{14}N, ^{13}C and secondary products like ^{20}Ne, ^{23}Na and ^{22}Ne;
- in the slightly deeper zone where helium burns, in particular during the thermal pulses, there is an important production of ^{12}C. During these pulses, the combustion zones of helium are in contact with regions enriched in ^{14}N, ^{13}C and other products

(Continued)

Box 4.1 (Continued)

generated in the hydrogen combustion layer, so that new isotopes are formed by capture of helium nuclei, for example ^{18}O from ^{14}N. The reaction $^{13}C + ^{4}He \rightarrow ^{16}O + n$ is of particular importance, because it produces neutrons which are at the origin of a very interesting nucleosynthesis, involving neutron capture by pre-existing nuclei followed by β^- disintegration when the synthetised products are unstable. The neutron flux is small enough for β^- disintegrations to occur before the capture of another neutron. The elements formed in this way are called *s elements* (*s* for slow), the neutron capture being designated as "slow". The reactions are of the general type:

$$(Z, A) + n \rightarrow (Z, A + 1) \rightarrow (Z + 1, A + 1) + e^- ,$$

Z and A being respectively the charge number and the atomic mass of the nucleus. It is in this way that Y, Sr, Ba, La, Nd, Tc, and so on until ^{208}Pb are formed. Technetium, whose isotopes, all radioactive, have periods equal to or shorter than 4.2×10^6 years (the period of ^{98}Tc), provided the first direct observational evidence for stellar nucleosynthesis, as ^{98}Tc has been observed in the spectrum of several AGB stars;

- finally, if the initial mass of the star is higher than 4 $M_\odot$, the temperature at the base of the convective envelope is high enough for hydrogen burning, the nucleosynthesis products being immediately carried to the surface by convection: these are amongst others ^{14}N, ^{26}Al and ^{7}Li; the latter results from decomposition of ^{7}Be produced by the proton–proton reactions.

The post-AGB evolution has only been observed for a few tens of years, thanks to radioastronomy and infrared observations from the ground or from space. In radio, we detect the molecules formed in the cold envelope, in particular CO in all the types of AGB stars and OH in those which are rich in oxygen. We call those AGB stars which are seen in the lines of OH and in the infrared but are not visible optically as *OH/IR stars*. The characteristic profile of the OH lines allows us to obtain the velocity of expansion of the envelope, a few tens of km/s. In the infrared, what is observed is the thermal emission of dust at the different stages of evolution, as shown in Figure 4.1.

4.1.2 Planetary nebulae

When the central star has become hot enough for its ultraviolet radiation to ionise the envelope, a bright planetary nebula appears around this star.

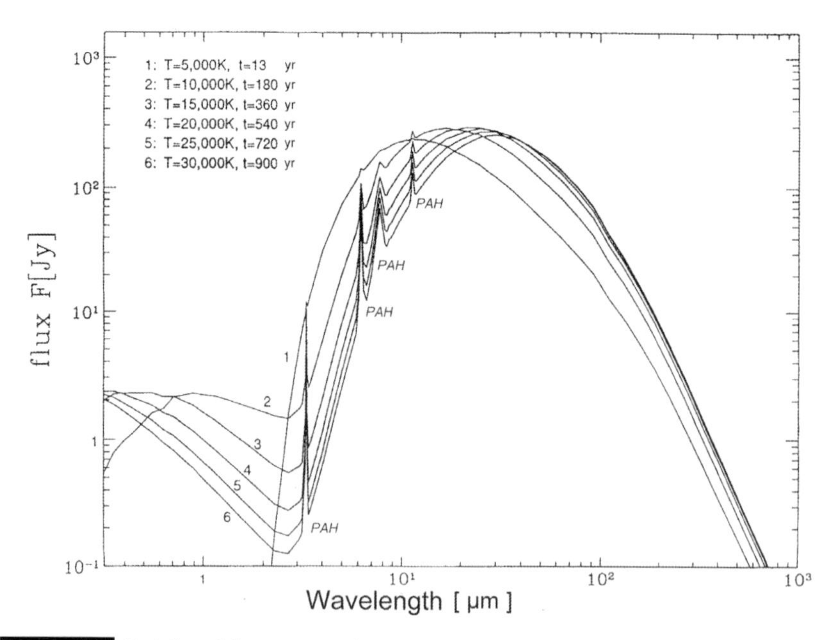

Figure 4.1. Evolution of the spectrum of a post-AGB star from the end of the TP-AGB stage to that of planetary nebula. The different steps with the surface temperature of the central object are indicated. The short-wavelength part of the spectrum is emitted by the central object, which is completely invisible optically at the beginning but becomes progressively visible while it heats up and the envelope is dispersed. The emission of the envelope is in the mean and far infrared. This envelope cools, the nature of the grains change and the characteristic broad emission lines of the polyaromatic ogenated (PAH) very small grains appear if the C/O ratio is larger than 1, as it is the case here. What is shown is the result of a numerical simulation, which is in excellent agreement with observations. *From Acker, A. in Lequeux et al. (2009).*

Its radiation is due to the recombination lines of hydrogen and to forbidden lines like those of OIII.[1] The name of "planetary nebula" is very improper as these objects have nothing to do with planets, but it survives. In 1764, Messier catalogued a hundred nebulae of unknown nature which looked in small telescopes like disks vaguely similar to planets; their name of planetary nebulae was given by Darque in 1779. Because these objects are very bright, they were amongst the first ones to be observed spectroscopically, in particular by William Huggins in the second half of the 19th century.

[1] However, it may be that the ejected matter is dispersed before the central star is hot enough to ionise it. In this case no planetary nebula is formed. This is often the case for stars of small masses.

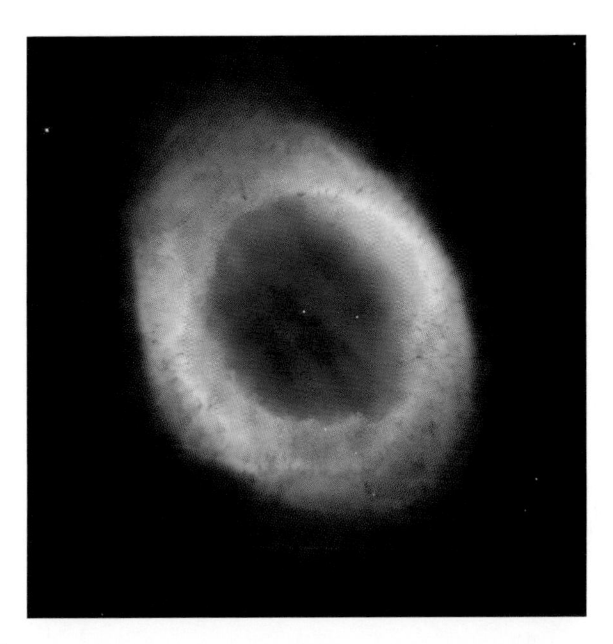

Figure 4.2. The planetary nebula M 57 observed with the Hubble Space Telescope. This nebula, ejected and ionised by the star visible at the centre, has a diameter of about 0.3 pc. The central star has an effective temperature of about 120 000 K and radiates essentially in the far ultraviolet, emitting little light in the visible. The colours are due to emission lines whose relative intensity varies with the distance to the star. © *Hubble Space Telescope Heritage.*

Some of these planetary nebulae are approximately circular or elliptic, which means they are probably ellipsoids, a sign that the wind that produced them was more or less isotropic. An example is shown in Figure 4.2. But these are special cases. The shape of planetary nebulae is very varied, and we can sometimes observe jets from the star; thus the wind can be very anisotropic. This may come from the fact that the central star is a binary. We often see several concentric envelopes, a sign that there were several successive ejections. The neutral material survives for some time around the ionised gas and can be observed through molecular lines like those of CO at 2.6 and 1.3 mm wavelength; we often observe vibration lines of H_2 in the near infrared that originate in the compressed interface surrounding the ionised gas, whose pressure is quite high. Examples are shown in Figures 4.3, 4.4 and 4.5.

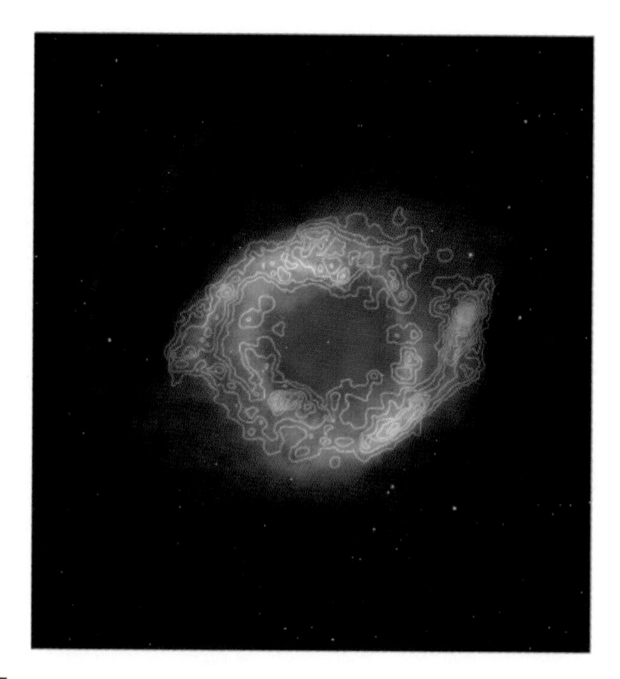

Figure 4.3. The Helix planetary nebula. The white contours show the distribution of the 1.3 mm line of the CO molecule. They are superimposed over an image obtained with the Hubble Space Telescope. Notice the central star. *From Young, K. et al. (1999) Astrophysical Journal 522, 387, with permission of the AAS.*

The chemical composition of the planetary nebula is that of the envelope of the star at the end of the TP-AGB phase. The C/O ratio can be either smaller or larger than 1. The central star is of Wolf–Rayet type in 10% of the cases: it is then surrounded by a residue of the envelope in fast expansion, which produces broad emission lines.

4.1.3 White dwarfs

Figure 4.6 shows the evolution of central stars of different masses. After an effective temperature of 100 000 to 400 000 K is reached depending on the mass, the luminosity of the star decreases rapidly and it becomes a white dwarf, whose matter is fully degenerate. There is no source of nuclear energy anymore, and the star can only cool. This cooling is rapid at the beginning as shown in Figure 4.6, and an effective temperature of

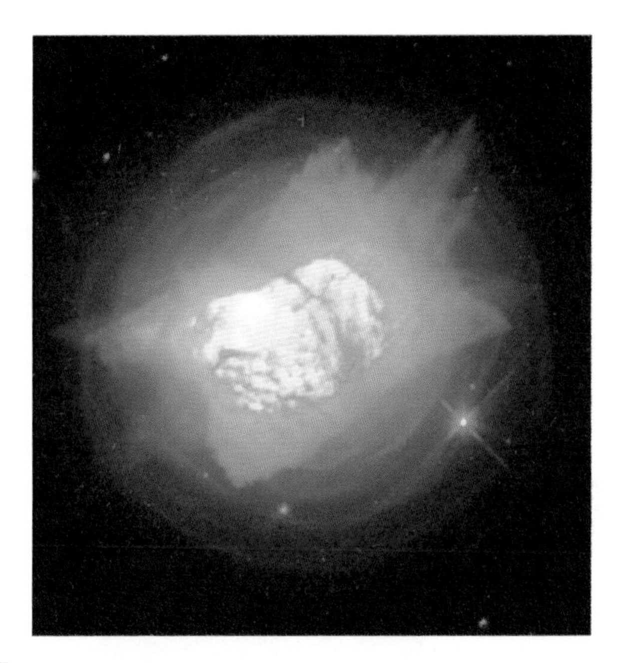

Figure 4.4. The planetary nebula NGC 7027, seen with the Hubble Space Telescope. This is a composite image resulting from the super position of an image on the visible (blue) over another one in the near infrared (yellow and red). Notice the external shells formed by several successive ejections. © *Hubble Space Telescope Heritage.*

the order of 10000 K is reached, for which the star has a white colour, hence its name. Then the cooling slows down considerably because the radiated energy is proportional to T_{eff}^4 and the surface of the star is very small, its radius being comparable to that of the Earth. It takes about 10^{10} years to reach an effective temperature of 4000 K, somewhat more for higher masses. As a consequence, most of the white dwarfs formed since the beginning of the Galaxy 1.4×10^{10} years ago are still observable, but only near the Sun because of their dimness. We know more than 3000 of them in the immediate vicinity of the Sun. They form a very numerous population which can give us information about the past rate of star formation in the Galaxy.

The spectra of white dwarfs is of interest for physicists: the very high gravity at their surface causes a considerable broadening of their spectral lines, and it has been possible to measure the gravitational redshift of these lines in spite of their broadness.

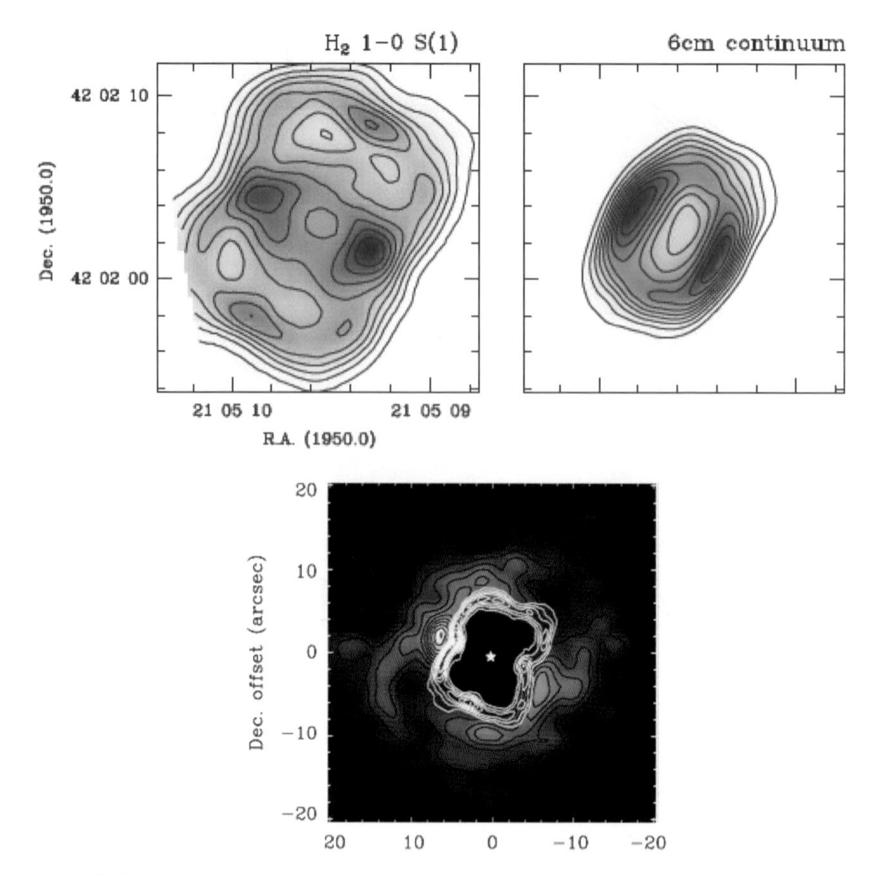

Figure 4.5. Several other aspects of the planetary nebula NGC 7027 (compare to Figure 4.4). Top, in a vibration line of molecular hydrogen at 2.12 μm and in the radio continuum at 6 cm, at the same scale. The radio continuum traces the ionised gas. Bottom, the red image with black contours depicts the distribution of the CO molecule, observed in the 2.6 mm line (different scale); the white contours show the exterior of the infrared emission of the molecular hydrogen: its vibration lines are emitted in a zone intermediate between the ionised gas and the molecular envelope. *From Cox, P. et al. (1997) Astronomy & Astrophysics 321, 907, and (2002) Astronomy & Astrophysics 384, 603, with permission of ESO.*

4.2 The death of big stars

4.2.1 Supernovae

Contrary to small stars, the stars with a mass larger than about 8 $M_\odot$ end their lives explosively as supernovae. The luminosity of supernovae can

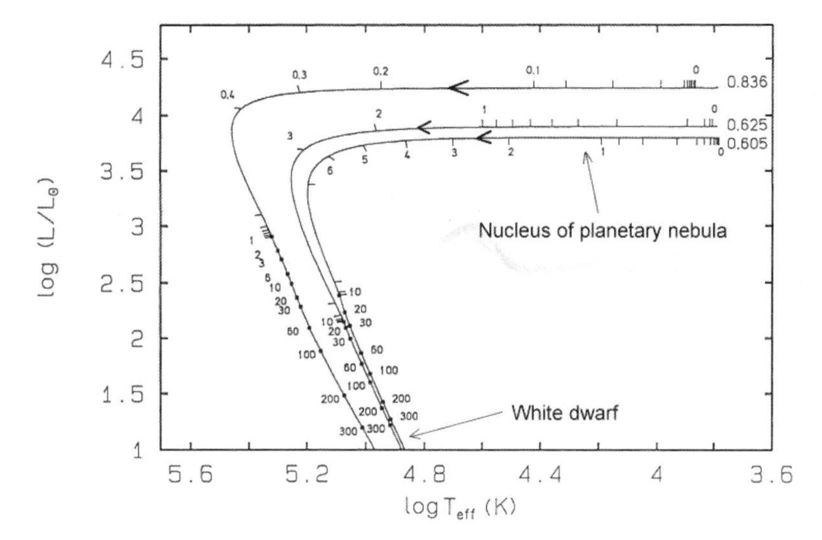

Figure 4.6. Evolutionary tracks of post-AGB stars in the HR diagram. The two lower tracks are for stars with initial masses 3 $M_\odot$ and core masses 0.605 and 0.625 $M_\odot$ respectively; the upper track is for a star with an initial mass 5 $M_\odot$ and a core mass 0.836 $M_\odot$. Time is given in thousands years along of each track. *From Blöcker, T. (1995) Astronomy & Astrophysics 299, 755, with permission of ESO.*

equal that of a whole galaxy for a few weeks (Figure 4.7). About ten supernovae have been seen with the naked eye since Antiquity, in particular one in 1054 by Chinese astronomers, then in 1572 by Tycho Brahe and in 1604 by Kepler. The remnants of these supernovae are known; we will describe them later. The last supernova observed in our Galaxy exploded about 350 years ago in the constellation Cassiopeia, in a direction where extinction of light by interstellar dust is very large. It is only known by its remnant, which is the brightest radio source in the sky. Flamsteed catalogued in 1680 a star which disappeared later, whose position is close to that of the remnant, but this could be an error. Several other supernovae have certainly exploded in our Galaxy since that time, but too far to be observed because of interstellar extinction. Very many supernovae have been discovered in external galaxies by systematic searches with cosmological purposes: probably more than 1 000 are known as of the beginning of 2013. On the average, 2 to 3 supernovae explode every year in an average galaxy like ours.

Figure 4.7. The supernova SN 1998bw in the distant spiral galaxy ESO 184-G82. Left, the galaxy before explosion (observation with the Schmidt telescope of ESO); right, after the explosion, the supernova is indicated by an arrow (observation with the 3.5 m New Technology Telescope of ESO, for which the images are sharper). At a distance of about 45 million parsecs, this supernova was at least as luminous as the rest of its parent galaxy: it is actually one of the most luminous supernovae ever observed. It results from the explosion of a very massive star — a hypernova — which produced simultaneously a gamma-ray burst. The name given to a supernova corresponds to the year of its discovery followed by letters indicating the order of its discovery in that year: this one is thus the 26 + 23 = 49th supernova discovered in 1998. © *ESO.*

There are supernovae of different types, reflecting the properties of the star at the time of explosion. They are classified mainly according to the appearance of their spectra. Type I supernovae (SN I) do not show hydrogen lines, while these lines are present in Type II supernovae (SN II). SN Ia is a subclass of SN I characterised by the presence of lines of silicon. The maximum luminosity of SN Ia and the shape of their light curve (the curve giving the luminosity as a function of time) appear very similar from object to object, so they are considered as "standard candles", which allow us to determine the distance of their parent galaxy. SN Ia do not result from the explosion of massive stars, but correspond to the thermonuclear explosion of a white dwarf in a close binary: we will describe this in the next chapter. The other supernovae (SN Ib, Ic, II, etc.), which are approximately five times more numerous than the SN Ia in a spiral galaxy like ours, all correspond to the explosion of massive stars. The considerable differences that exist between these supernovae result from the details of the explosion, in particular from the chemical composition of the envelope at this time.

4.2.2 The race to the chasm

Let us consider the star as we left it in Section 3.4. It consists in a core of carbon 12 and oxygen 16 resulting from the previous combustion of helium, surrounded by a layer where helium 4 burns. If the mass loss was moderate, another layer exists at a larger radius where hydrogen burns, and there is an enormous envelope defining a supergiant. If mass loss was large, as in Figure 3.11, this envelope and even the hydrogen-burning shell do not exist, and the star is a Wolf–Rayet. All this has no relevance for the moment: what is important is what occurs in the core. As there are no nuclear reactions in the core, it can only contract and its temperature rises. When it reaches 6 to 8 × 10^8 K, carbon starts to burn. The temperature continues to rise and when it reaches 1.9 × 10^9 K, hydrogen fusion starts after a complex episode in which the fusion products of carbon intervene, in particular ^{20}Ne. The fusion of ^{16}O synthetizes, amongst various products, several isotopes of silicon, especially ^{28}Si. At 3.3 × 10^9 K, photodisintegration reactions occur, producing helium nuclei α, for example:

$$^{28}\mathrm{Si} + \gamma \rightarrow {}^{24}\mathrm{Mg} + \alpha.$$

The α produced in this way are captured by various nuclei, producing heavier and heavier nuclei. The most stable survive. We go in this way until ^{56}Fe, which is the most stable nucleus of all. Beyond ^{56}Fe, other nuclei can be produced, but only by neutron capture. The star has now an onion shell structure, which is depicted schematically in Figure 4.8. The matter is at the limit of degeneracy, and eventually becomes degenerate in the central regions at the end of this evolution, at a radius which depends on mass and rotation.

During all these phases, neutrinos are produced in increasing quantities while temperature is rising. They come mainly from the annihilation of electron–positron pairs, themselves produced by materialisation of photons with energy larger than 0.511 MeV: $\gamma + \gamma \leftrightarrow e^- + e^+$. These photons form

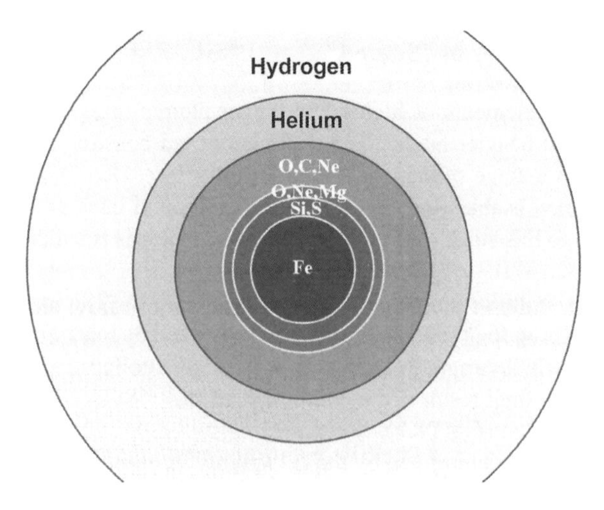

Figure 4.8. Scheme of the layer structure of a massive star just before it explodes. *From Maeder (2009).*

the tail of the energy distribution and are abundant when the temperature rises above the fusion temperature for carbon (an energy of $E = 0.511$ MeV corresponds to a temperature $T = E/\mathrm{k} = 5.93 \times 10^9$ K). The annihilation of an e^-–e^+ pair can produce either a pair of photons or an electronic neutrino–antineutrino pair, the ratio of the respective probabilities being:

$$P(\nu)/P(\gamma) = 3 \times 10^{-18}\,(E_\nu/\mathrm{m_e c^2})^4,$$

where E_ν is the energy produced in the annihilation: $P(\nu)$ is thus a fast function of temperature.

The neutrinos escape from the star without problem, carrying energy with them. In order to compensate for this loss, the star contracts, heats up and burns its fuel in an unrestrained way without reaching stability. Conversely, the higher the temperature, the faster the nuclear reactions and production of energy and neutrinos: we race to a catastrophe. Moreover, when the temperature reaches 10^{10} K and the density 10^{10} g cm^{-3} in the core, the photons have enough energy to photodisintegrate iron through an endothermic reaction which uses a large quantity of thermal energy, accelerating the collapse:

$$^{56}\mathrm{Fe} + \gamma \rightarrow 13\alpha + 4\mathrm{n} - 124\ \mathrm{MeV}.$$

The duration of the different phases of the combustion and the corresponding physical conditions are given in Table 4.1. Finally, Box 4.2 gives details on the nuclear reactions and nucleosynthesis in the core.

Phase	Duration	Temperature	Specific mass in the core	Principal combustion products
H burning	7×10^6 years	6×10^7 K	5 g cm^{-3}	^{4}He, ^{14}N
^{4}He burning	5×10^5 years	2×10^8 K	700 g cm^{-3}	^{12}C, ^{16}O
^{12}C burning	600 years	9×10^8 K	2×10^5 g cm^{-3}	^{24}Mg, ^{20}Ne
Photodisintegration of ^{20}Ne	0.5 year	1.7×10^9 K	4×10^6 g cm^{-3}	^{16}O, ^{24}Mg, ^{28}Si
^{16}O burning	6 days	2.3×10^9 K	10^7 g cm^{-3}	Isotopes of Si and S
Photodisintegration of ^{28}Si	1 day	4.0×10^9 K	3×10^7 g cm^{-3}	^{56}Fe, etc.

Table 4.1. Combustion phases in the core of a 25 $M_\odot$ star. Their characteristics are only approximate because they depend on the rotation of the core.

Box 4.2. Nucleosynthesis in massive stars

At the end of the life of massive stars, their core is initially made of ^{12}C and ^{16}O, which result from the fusion of helium. Then, as the temperature rises due to gravitational contraction, fusion of carbon occurs, producing an excited nucleus of ^{24}Mg* which disintegrates in several ways, the most important being:

$$^{12}C + {}^{12}C \rightarrow {}^{24}Mg^* \rightarrow {}^{20}Ne + {}^4He\ (\alpha)$$

$$\rightarrow {}^{23}Na + {}^1H\ (p).$$

Various reactions convert ^{23}Na into ^{20}Ne, and a part of the ^{20}Ne into ^{24}Mg. The main ones are:

$$^{23}Na + p \rightarrow {}^{20}Ne + \alpha$$

$$^{20}Ne + \alpha \rightarrow {}^{24}Mg + \gamma.$$

At the end of carbon burning, the main products are therefore ^{20}Ne and ^{24}Mg, while a large fraction of ^{16}O remains runtouched.

The next episode is the photodisintegration of ^{20}Ne, when the energy of photons is large enough due to the rise in the temperature:

$$^{20}Ne + \gamma \rightarrow {}^{16}O + \alpha,$$

the α being captured by the remaining ^{20}Ne:

$$^{20}Ne + \alpha \rightarrow {}^{24}Mg + \gamma.$$

Another important reaction with the α is:

$$^{24}Mg + \alpha \rightarrow {}^{28}Si + \gamma.$$

Oxygen fusion starts at 1.9×10^9 K: it produces $^{32}S^*$ whose disintegration gives mainly ^{28}Si and ^{32}S, while other reactions produce various isotopes of Si, P, S, Cl, Ar, K, Ca and Sc.

Fusion of ^{28}Si does not occur, because it would require a very high temperature before which various photodisintegrations occur such as:

$^{28}Si + \gamma \rightarrow {}^{24}Mg + \alpha$, and similar reactions with ^{24}Mg, ^{20}Ne, etc.

Reaction chains produce heavier and heavier elements by capture of the α liberated by the preceding reactions. The most stable of these elements subsist, until ^{56}Fe which is produced in large quantities.

Other elements, in particular those heavier than iron, are synthetised during the supernova explosion itself: see Box 4.3.

4.2.3 The explosion

At the end of the episode described in the preceding section, the iron core is degenerate. When its mass exceeds the Chandrasekhar limit (see Section 2.6), the Fermi pressure of electrons cannot fight against gravity and the core collapses. The gravitational energy liberated in this collapse is enormous, several 10^{53} ergs, over a very short time, about a hundredth of a second. Most of this energy, about 99%, is produced as neutrinos, and the remaining energy, about 10^{51} ergs, as kinetic energy. Of these 10^{51} ergs, 1% is converted into luminous energy; therefore, although the appearance of a supernova is a spectacular event, it represents only 1/10 000 of the total liberated energy.

A problem is that we do not really understand how the kinetic energy of the collapse is transferred to the external, non-degenerate layers. These layers do not immediately realise that "the ground has collapsed under their feet", so that their evolution is completely decoupled from that of the collapsing central parts. During the core collapse, the core matter reaches specific masses so high that most of the electrons are captured by nuclei and protons, which are transformed into neutrons. This neutronisation is accompanied by the emission of neutrinos (one per synthetised neutron), which try to find a way out. However, the density of matter in the core is now so large that they cannot escape immediately: this matter has become opaque to neutrinos, as it was before to photons. When the density at the centre of the core reaches 2×10^{13} g cm^{-3}, the density of nuclear matter, the collapse is stopped. This generates gravity waves: the core starts to vibrate. These vibrations carry to the rest of the star a part of the gravitational energy. But they encounter matter which is still collapsing. A fight occurs between the wave coming from inside and the avalanche falling from outside. It was hoped that the wave would win, but in vain: present numerical simulations show that the wave is not powerful enough to trigger the explosion of the external layers, so that the star collapses entirely into a black hole. Some scientists actually believe that such a total disappearance could be the case for very large mass stars.

Let us now consider the neutrinos. They are prisoners only for a short time and eventually reach the surface of the core as an enormous burst of 10^{58} neutrinos, each one with an energy of 15 MeV corresponding to the core temperature of about 1.5×10^{11} K, the highest temperature known in the present Universe. 1% of these neutrinos interact with the protons and neutrons outside the core, giving them their energy, somewhat more than 10^{51} ergs. Numerical simulations show that this delayed heating could trigger an explosion, but only for stars whose initial mass is smaller than 11 $M_{\odot}$. For higher masses, this mechanism seems insufficient. However, we do not really know what the mass limit is, or even if such a mass exists.

As astrophysicists are never short of imagination, they have thought about the possibility that various hydrodynamic instabilities could generate the explosion via the intermediary acoustic waves generated by the gravity waves. Some numerical simulations of these instabilities are promising and predict an asymmetry of the explosion which would have a preferred direction. This could explain the fact that the neutron stars which are the remnants of the nuclei of some massive supernovae are often ejected with high velocities. However, we must acknowledge that, in spite of some promising tracks explored over the last few years, the mystery of the explosion of massive supernovae has not been solved.

Whatever its cause, the explosion is the seat of very active nucleosynthesis dominated by the neutron capture of different nuclei. It is schematised in Box 4.3.

Box 4.3. Nucleosynthesis in supernova explosions

During the core collapse of a supernova, many neutrons are present in the outer zones. They can be captured by nuclei. The neutron density is between 10^{19} and 10^{25} neutron/cm^3, so that the interval between two successive captures of a neutron by a nucleus is between 10^{-6} and 1 second, given the order of magnitude of capture cross sections. An equilibrium sets in between neutron capture and photodisintegration of the formed nuclei:

$$n + (Z, A) \leftrightarrow (Z, A+1) + \gamma,$$

where Z and A are respectively the charge number and the atomic mass of the nucleus. This is the *r process* (r for rapid), which contrasts with the s process (Box 4.1) in which the nuclei formed by neutron capture have time to disintegrate either by photodisintegration or by simple radioactive disintegration before capturing another neutron. The r process forms heavier and heavier nuclei, possibly through unstable nuclei if their lifetime against β^- decay is longer than the interval between successive neutron captures. During this process, there is an accumulation of nuclei with magic numbers of protons and neutrons, for which the cross section for neutron capture is particularly small. When the irradiation by neutrons is finished, the unstable elements disintegrate until stable nuclei are formed. For the heaviest nuclei, fission can occur and it is these fissions which determine the heaviest nuclei formed by the r process: we think that this occurs for unstable nuclei beyond uranium, so that uranium is the heaviest nucleus synthetised through the r process.

The r elements tend to be more neutron rich than the s elements, and can be distinguished in this way. For example, under the conditions of the s process, the radioactive isotope ${}^{115}_{48}$Cd produced by neutron capture decays via β^- emission in 54 hours to form the stable isotope ${}^{115}_{49}$In, while under the conditions of the r process ${}^{115}_{48}$Cd captures a neutron before decaying to become ${}^{116}_{48}$Cd.

It is possible to reproduce rather well the observed abundances of the r elements with a temperature $T \geq 10^9$ K and a neutron density of the order of 10^{20} cm^{-3}, but this does not give us strong constraints on the place where the r process occurs during the explosion, which remains hypothetical.

Massive supernovae are probably also at the origin of some rare neutron-poor elements. These are the *p elements* (p for poor). They could be synthetised during the explosion from pre-existing s elements, via (γ, n) reactions in O/Ne layers at temperatures of the order of 3×10^9 K.

4.2.4 SN 1987A

Although we do not understand well the mechanism of explosion of supernovae, one thing is sure: they do explode. The neutrinos produced during the explosion have indeed been detected, as we will see in the example of SN 1987A.

This supernova appeared on 23 February 1987 in the Large Magellanic Cloud, a satellite galaxy of ours, located at a distance of about 50 000 parsecs (Figure 4.9). Of course, all the telescopes in the Southern hemisphere looked at the supernova because this event was really exceptional. The results turned out to be as interesting as promised. However, while some observations did confirm the ideas of theoreticians, others posed serious problems which are not yet completely solved.

The progenitor of the supernova was identified, the first for any supernova. Although this star named Sk–69°202 had not been particularly studied before explosion, it was sufficiently well known: this was a blue

Figure 4.9. The supernova SN1987A in the Large Magellanic Cloud. Right, before explosion: the progenitor star Sk-69°202 is indicated by an arrow. Left, the supernova after its discovery. The cross is due to diffraction on the support of the secondary mirror of the telescope. © *Anglo-Australian Observatory.*

supergiant with an initial mass of about 20 $M_{\odot}$. At the time, we believed that the explosions occurred at the stage of red supergiant. This was not the case for Sk–69°202: the mass loss had been strong enough for the star to leave this stage and move in the HR diagram towards the region of Wolf–Rayet stars. Since then, it has been noticed that many supernovae explode as Wolf–Rayet stars: these are the Ib and Ic ones.

The most interesting result about SN 1987A, which opened a new area in astronomical observation, was the discovery of the arrival of about 20 neutrinos two and a half hours before the appearance of the supernova in visible light. This arrival was observed simultaneously in Japan with the Kamiokande neutrino detector and in the USA with the IMB one. It gave for the first time direct information on what occurs inside a supernova. The energy of the detected neutrinos, between 10 and 30 MeV, implies that the core temperature at the time of explosion was more than 10^{11} K, in agreement with the theory. From the number and energy of these neutrinos, which carry with them as we have seen almost all the energy produced by the core collapse, we deduced that the core collapse energy was 3×10^{53} ergs. For a few seconds, the supernova has thus emitted as much energy as all the stars of the observable Universe combined. However, not enough neutrinos have been observed to obtain information about the nature of the explosion. As to the delay of two and a half hours, this is the time that the shock wave produced at the centre took to arrive at the surface of the star.

Another very important result is the direct detection of radioactive nuclei in SN 1987A. ^{56}Ni, produced in the interior of the star as we will see later, disintegrated rapidly into ^{56}Co, which itself disintegrated more slowly into ^{56}Fe while emitting gamma photons with energies 847 keV and 1 238 keV. Photons with precisely these energies were detected by the American satellite SMM six months after the explosion. We could then confirm in this way that the energy liberated in the disintegration chain $^{56}\text{Ni} \rightarrow {}^{56}\text{Co} \rightarrow {}^{56}\text{Fe}$ feeds in a late phase the luminosity of massive supernovae. Previously, there was only an indirect evidence for this mechanism: the decrease of the luminosity follows the decay of ^{56}Co.

Finally, we could follow the fate of the matter ejected by the explosion and observe its interaction with the surrounding interstellar medium. One year after the explosion, an increase in extinction in the direction of the supernova showed that dust had been formed in the ejected matter. The supernova illuminated the neighbouring matter after a delay which depended on the distance of this matter to the supernova, and also to the Earth, due to the finite velocity of light (Figure 4.10). We could also see a ring of matter, with a radius of about one light-year, illuminated by the explosion surrounding what remained of the supernova. There is no doubt that this matter is made of circumstellar gas ejected by the progenitor before exploding. The fact that this is a ring and not a shell shows that the star was a rather rapid rotator. Later, two other rings parallel to the first one were also observed, probably resulting from previous ejections. The matter ejected by the explosion reached the first ring after 10 years, increasing considerably the luminosity of the ring which presently emits

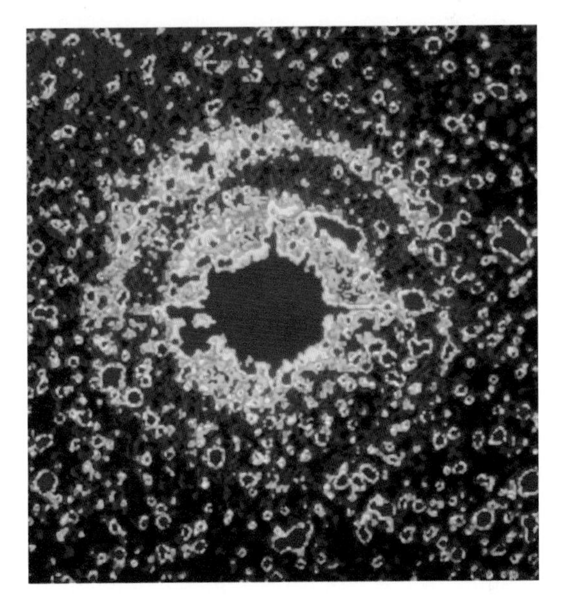

Figure 4.10. Interstellar matter illuminated by SN 1987A. This image was obtained one year after the explosion with the 3.60 m telescope of ESO. It is processed so that the weak light from the region around the very bright supernova is visible: the false colours indicate the intensity. The rings are the places where the matter was illuminated and diffused the light to the Earth at the time of the observation (plus of course the time that the light took to arrive from the Large Magellanic Cloud to us, about 160 000 years). There are clearly two layers of interstellar matter, that which produces the inner ring being closer to the supernova. A study of the 3-D distribution of interstellar matter is possible from these data and similar data taken at different times. © *ESO.*

Figure 4.11. The rings around SN 1987A, seen in 2010 with the Hubble Space Telescope. The bright ring surrounds what remains of the supernova, which is now quite weak. It is made of matter expelled by the progenitor star before the explosion. This matter is excited by the shock wave resulting from the arrival of the material ejected by the supernova explosion with a speed of about 30 000 km/s. We see two fainter rings, also expelled by the progenitor but which the ejected matter has not yet had time to reach. © *Hubble Space Telescope Heritage.*

X rays and radio waves, generated by a shock wave in the material of the ring. This shows that matter was ejected by the supernova with a speed of about 30 000 km s^{-1}. The excited ring looks like a pearl necklace, probably due to instabilities (Figure 4.11). Within a few more years, the material of the supernova will reach the two other rings. It remains to discover the central object, residue of the explosion, which can be either a neutron star or a black hole.

4.2.5 Hypernovae and gamma-ray bursts

During the late 1990s, astronomers realised that some SN Ic, which result from the explosion of extreme WC or WO Wolf–Rayet stars, had uncommon characteristics: their luminosity and ejection velocity are considerably larger than those of more common supernovae. A large amount of ^{56}Ni is observed spectroscopically in these objects. SN 1998bw (Figure 4.7) is an example. Given the large difficulties encountered when attempting to explain ordinary supernova explosions, how could we explain such monsters? By rotation. Without rotation, stars with initial masses larger than 25 $M_{\odot}$ collapse into a black hole and produce little ^{56}Ni. If the core rotates rapidly, there is still a black hole at the centre, but the outer parts of the core and the external layers cannot collapse directly into it and form a surrounding rotating disk. The viscosity in this disk, which is in differential rotation, produces a rapid loss of energy and the progressive fall (in a few seconds!) of the matter into the black hole. The amount of energy produced in this way is enormous. Like in protostars, an important part of this energy is released as two symmetrical jets aligned on the rotation axis of the disk. At their base, the temperature is several billion degrees, and a large amount of ^{56}Ni is produced by neutron capture from ^{56}Fe. The jets send matter in space with a velocity of the order of 30 000 km/s, while the rest of the matter is engulfed into the black hole, hence the name of *collapsar* which is sometimes given to these *hypernovae*. Their formation seems to only be possible when the metallicity is small, so that they are only observed in distant galaxies which are much less evolved, and thus less rich in heavy elements, compared to ours or other galaxies in our neighbourhood.

The hypernovae are probably the origin of a part of the gamma-ray bursts. These bursts were discovered by American military satellites in the 1960s; the purpose of these satellites was to check if the treatise of 1964, which banned nuclear tests in the atmosphere or in space, was respected: these explosions would have generated a large quantities of easily detectable gamma rays. Instead of this, the satellites observed bursts of gamma rays which could not originate in this way. This discovery was only made public in 1973. After this date, many observations were performed by satellites or space probes, but without solving the mystery. The situation evolved in 1991, when the NASA Gamma Ray Observatory (GRO) showed that the gamma-ray bursts are evenly distributed in the sky. They were thus either very close, or very distant: if not, they would have had a distribution more or less similar to that of Galactic stars. The question was definitively settled by the Italian–Dutch satellite Beppo-SAX, launched in 1996: it allowed for the first time the position of a gamma-ray burst, which was accompanied by a X-ray burst much easier to localise, to be found accurately. The information on the position was immediately transmitted to several observatories succeeded in identifying an optical counterpart: the burst came from a very distant galaxy, with a redshift corresponding to a distance of 9 billion light years. Since that time, many host galaxies of gamma-ray bursts have been identified. They are all very distant.

Given these distances, the gamma-ray bursts are necessarily very luminous. If the emission was isotropic, it would correspond to energies of more than 10^{53} ergs as gamma-rays alone, which is close to the total energy produced by a typical supernova. However, it seems completely impossible to convert all the energy into gamma photons: we already saw that 99% is dissipated as neutrinos. Thus the burst is probably very anisotropic, and we see it only when it is directed toward us. Even in this case, the energy emitted as gamma rays must be very high, of the order of 10^{51} ergs, to be compared to the 10^{49} ergs of the electromagnetic emission of usual supernovae. We think that the symmetrical jets emitted by the hypernovae contain relativistic protons and electrons, launched in the jet with different velocities. Their collisions would produce gamma photons directed along the jet (Figure 4.12).

We can also hypothesise that a fraction of the gamma-ray bursts come from the fusion of two neutron stars in close binary systems: we will discuss this in the next chapter. The present consensus is that hypernovae account for the longer gamma-ray bursts (a few minutes) whilst the second process explains the shorter bursts which last for about one second.

4.2.6 The supernovae from e^+–e^- pair instability

We saw that when the central temperature of the core is well above 10^9 K, photons begin to materialise into electron–positron pairs. The disappearance of photons, whose radiation pressure was important, decreases the adiabatic exponent γ to less than 4/3, favouring the instability of the core. If more than 30% of the mass of the core is affected by this instability, which is the case for initial masses larger than about 100 $M_\odot$, a collapse occurs at the time of ^{20}Ne disintegration. Then two cases are possible:

- either the pressure increase due to the collapse re-ignites the fusion of oxygen, so that the energy produced by this fusion stops the

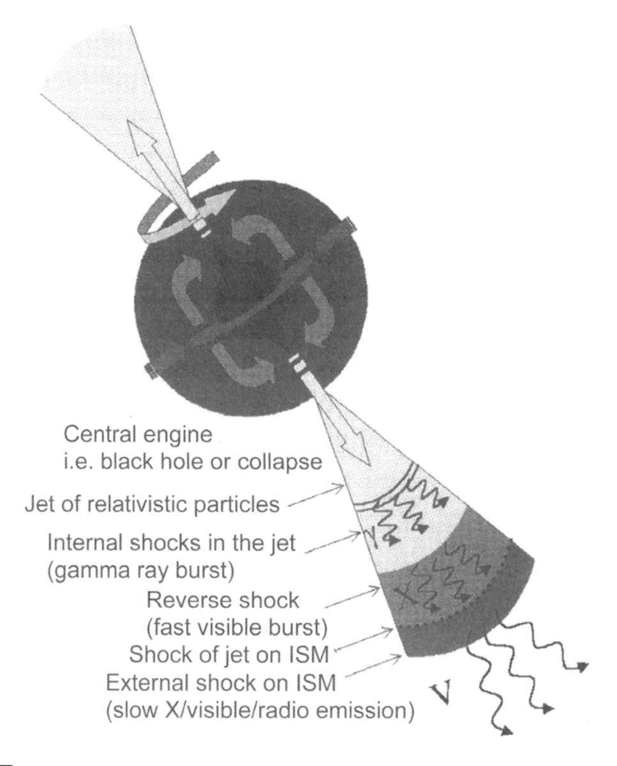

Figure 4.12. Scheme of the generation of a gamma-ray burst by a hypernova. The gammas are generated by collisions of charged particles in inner shocks, while the longer wavelengths (X, visible, radio) come from an outer shock. *From Prantzos, N., in Lequeux et al. (2009).*

contraction while a part of the envelope is expelled. The cycle repeats until the star has lost enough mass to explode as an ordinary supernova;

- or the star has a really large initial mass, more than 130 $M_\odot$, and the collapse of the core, whose mass is of the order of 50 $M_\odot$, is so fast that even oxygen fusion cannot stop it. The energy produced is now so large that the star explodes entirely, without leaving a compact remnant. We speak in this case of a pair-instability supernova.

These conclusions are somewhat speculative, but it seems that an extremely luminous SN Ic, SN 2007bi, belongs to this category.

4.2.7 Supernova remnants

What is left of a supernova after its explosion? In general, a condensed object which can be a neutron star of a black hole is left, as well as material ejected more or less isotropically. The Chandrasekhar mass limit determines the nature of the compact object: if the mass of the iron core is larger than about 2 $M_{\odot}$, it is too large to form a neutron star, and the residue is a black hole. If there were no mass loss, the limit would correspond to a stellar mass of 25 $M_{\odot}$. Mass loss and rotational mixing raise this initial mass to 40 to 80 $M_{\odot}$. For an initial mass of up to about 130 $M_{\odot}$, the central object is a black hole; for even greater masses, there is no compact remnant at all, as we have seen.

In many cases the remaining neutron star has been detected as a pulsar. In effect, neutron stars generally rotate very fast, and the huge magnetic field resulting from the compression of the conducting matter, which is often of the order of 10^{12} gauss (10^{8} tesla), accelerates the matter up to relativistic energies in two opposed jets. In these jets, a complex mechanism which has not been entirely elucidated produces a radio emission in their direction. If this direction crosses the observer, he receives radio pulses. When the residue of the explosion is a black hole, it is almost impossible to detect. However, we will see in the next chapter that one of the components of some close, massive binary stars is a black hole, resulting probably from the explosion of a progenitor as a supernova.

The ejected matter forms a supernova remnant, which survives a few ten thousand years before it disperses into the interstellar medium. There are about 250 such remnants in the Galaxy, in agreement with the explosion rate of supernovae and the lifetime of these remnants. The appearances of these remnants differ depending on there is a pulsar inside or not. When there is no pulsar, the remnant is not fed in energy anymore and looks like an apparently empty expanding shell whose size depends on its age (Figure 4.13). If there is a pulsar, it feeds the remnant with high-energy particles accelerated in its jets: the remnant is filled with these relativistic particles, especially electrons which radiate through the synchrotron mechanism by interacting with the magnetic field (Figure 4.14): we then speak of a *plerion* (from the ancient Greek *pleres*, full), or of a PWN (pulsar with nebula). The ejected matter can form a shell in a later phase of evolution, so that there are intermediates between the two basic shapes.

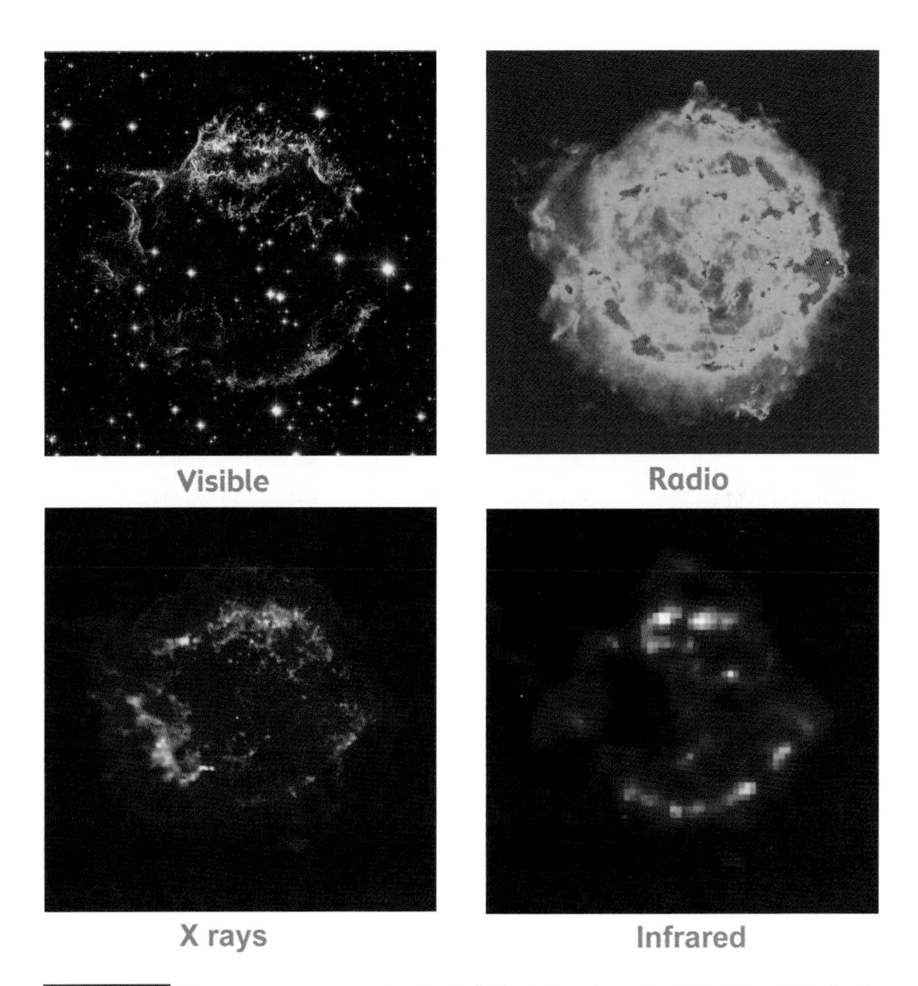

Figure 4.13. The supernova remnant Cassiopeia A. Top left, an image in visible light obtained with the Hubble Space Telescope, covering 5 × 5 parsecs. The emission is from the ejected matter and interstellar matter compressed by the shock wave. Top right, a radio image obtained with the Very Large Array; the emission is the synchrotron radiation of electrons accelerated in the shock in the magnetic field of the compressed matter. Bottom left, an image in X rays obtained with the satellite CHANDRA; here, we observe the thermal emission of the compressed gas, whose temperature is very high. The central object could be a neutron star, residue of the explosion. Bottom right, an image in the mean infrared (10.7 to 12 μm) obtained with the ISO satellite. It shows the thermal emission of dust condensed in the gas ejected by the explosion. All images are at the same scale. © *NASA, National Radio Astronomy Observatory and European Space Agency.*

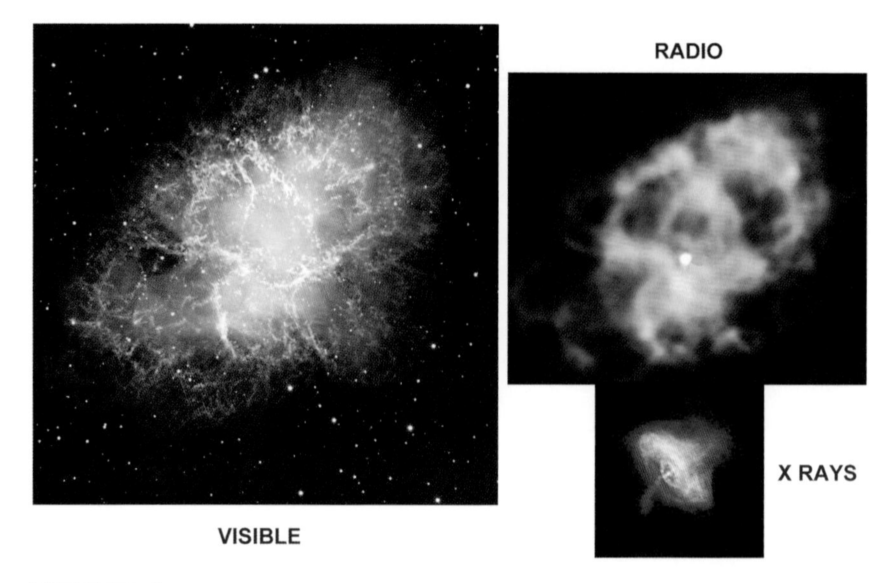

Figure 4.14. The Crab nebula. This is a typical plerion, continuously fed with high-energy particles by a central pulsar. Left, an image in the visible covering 4 × 4 parsecs, obtained with one of the 8-m telescopes of the Very Large Telescope of ESO. The filaments are remnants of the ejected matter and form a shell in fast expansion. The blue, diffuse light is the synchrotron emission by relativistic electrons accelerated by the pulsar in the magnetic field of the object. Top right, the synchrotron radio emission of these electrons, obtained with the Very Large Array; notice the central pulsar. Bottom right, the synchrotron emission in X rays of the most energetic electrons, with a morphology different from that at longer wavelengths (image from the CHANDRA satellite). Notice the jets emitted more or less symmetrically by the pulsar. All three images have the same scale. © *ESO, National Radio Astronomy Observatory and NASA/Harvard-Smithsonian Center for Astrophysics.*

4.2.8 The runaway stars

It might be that, in a binary star formed of two massive components, one explodes as a supernova. Just after the explosion, a large amount of mass is ejected and a low-mass compact object is left. The other component may not remain gravitationally tied to this object and then the binary system disperses. The massive star leaves with a velocity slightly smaller than its former orbital velocity, which is in general about several tens of km/s. The same occurs to the compact remnant, a neutron star or black hole. We observe that an important fraction of O and B stars have a large velocity: they are called the *runaway stars*. Many pulsars also have large velocities: these are neutron stars, residues of supernova explosions.

5

The Zoo of Double Stars

Many stars do not seem to conform to the evolutionary scheme described in the preceding chapters. This is the case for some supernovae, for novae (which are stars experiencing less powerful explosions than supernovae) for stars which are strong X-ray sources, etc. All these objects are actually close binary stars whose evolution has been strongly modified by the interaction between the components. There are many other stars which belong to double or triple systems, but which are more distant from each other, so that they can evolve independently like isolated stars. More than half of all stars belong to double or multiple systems, close or not.

Research on interactive binaries started in the 1960s. Their importance has grown continuously since and led to numerous surprises, often spectacular. This explains why we devote to close binaries an entire chapter of this book. Before describing the various aspects of what a true zoo is, let us recap how the principal physical parameters of stars can be determined from observations of binaries.

5.1 Double stars and stellar masses

As we said in Chapter 1, observation of double stars is the only way to obtain the masses of stars directly. In some cases, it also makes it possible to determine their radius and atmospheric parameters. It is for this reason that, from the beginning of the XIXe century, the observation of *visual binary stars*, i.e. those which can be seen separately by telescopes, was a major topic of observational astronomy. The epoch of professional or amateur astronomers measuring the separation and the orientation of these binaries visually is over; it is possible today to obtain the same result with better precision using more refined methods: *speckle interferometry* and *adaptive optics* allow us to benefit fully from the separating power of a large telescope in spite of atmospheric turbulence, and *interferometry* between several telescopes. On the other hand, spectroscopy can yield the time variations of the velocity with respect to the observer (the *radial velocity*) of one or both components of the system: the objects for which this is possible are the *spectroscopic binaries*. Finally, if we are lucky enough to observe eclipses of one of the components by the other, we know that we are located approximately in the plane of the orbit of the system. For these *eclipsing binaries*, we can obtain the radius and some atmospheric parameters of each component from a detailed study of the light curve, i.e. the time variation of the light received from the system. It is often necessary to be very patient in such studies, because the period of revolution of the system can be very long.

Let us examine now these different cases.

5.1.1 Visual binaries

If the two components of a binary star can be observed separately and if their relative motion can be followed, the orbit of one component around the other can be determined (it is not possible in general to know the location of the gravity centre, and hence to determine the motion of the components around it). What can be observed is the apparent orbit, which is the projection of the real orbit onto the plane of the sky. Figure 5.1 shows the geometry of the system.

Figure 5.2 shows how elements of the real orbit can be obtained once the apparent orbit is determined. We can demonstrate the following relations between the parameters of the true elliptical orbit (a = semi major axis; e = eccentricity; i = inclination of the orbit on the plane of the sky)

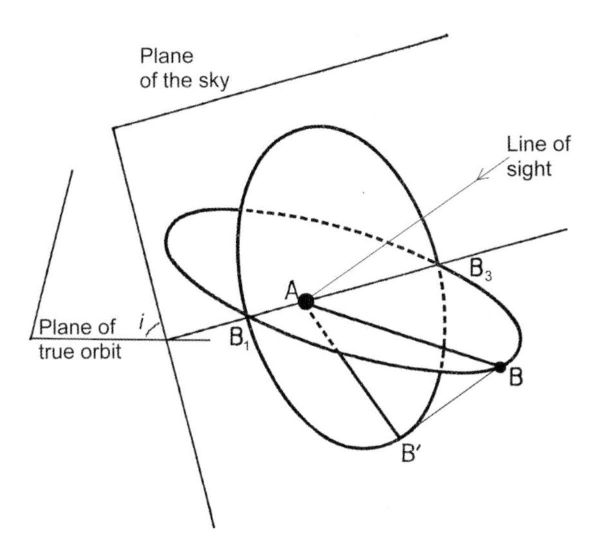

Figure 5.1. The true and projected orbit of component B of a visual binary around component A. The nodes B_1 and B_3 are on the intersection of the plane of the true orbit and the plane of the sky, which is by definition perpendicular to the line of sight, and over which the orbit is seen in projection. *i* is the inclination of the orbit.

and the lengths of the different segments depicted on the figure. We use as intermediates the quantities ρ_1, ρ_2, τ_1 and τ_2 defined as follows: $\rho_1 = AB_1 \times AB_3/B_1B_3$, $\rho_2 = AB_2 \times AB_4/B_2B_4$, $\tau_1 = (AB_3 - AB_1)/B_1B_3$, $\tau_1 = (AB_4 - AB_2)/B_4B_2$. We have:

$$e = CA/CP,\ \cos i = \rho_2/\rho_1\ ,\ a = (1 - e^2)/2\ \rho_2. \qquad (5.1)$$

Of course, the distance must be known in order to obtain the linear values of the semi major axis.

We can now derive the masses from the semi major axis *a* of the orbit and from its period *P* using Kepler's third law, which can be written as:

$$a^3/P^2 = G(M_1 + M_2)/4\pi^2. \qquad (5.2)$$

It is only possible in this case to obtain the sum of the masses M_1 and M_2 of the components. However, if we possess many determinations of

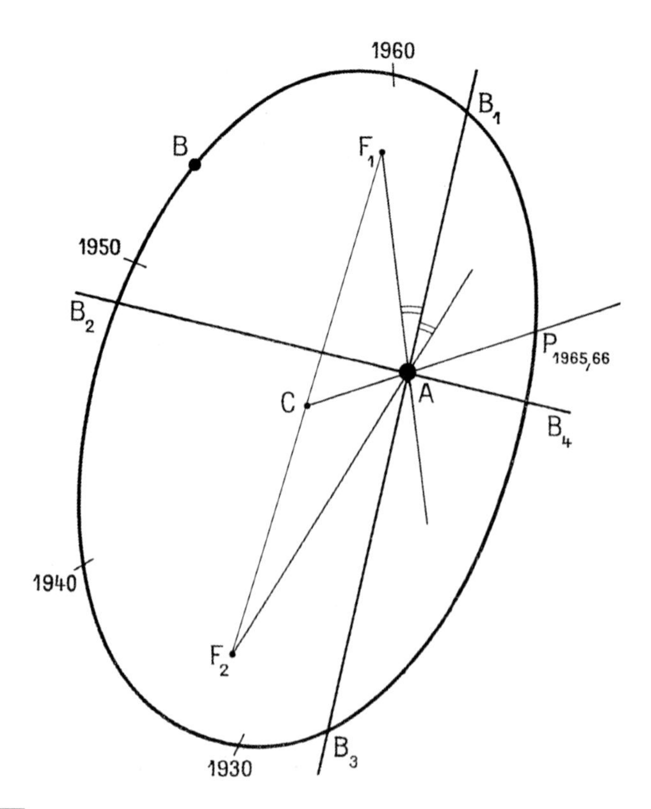

Figure 5.2. The apparent orbit of the companion of the star α Ursae Majoris. This star is at A; C is the centre of the apparent orbit, F_1 and F_2 its two foci. B_1B_3, parallel to F_1F_2, is the line of nodes, the intersection of the plane of the real orbit with the plane of the sky which contains the apparent orbit. B_2B_4 is perpendicular to B_1B_3. The positions of the companion at different epochs are indicated, and also the periaster P. *From Couteau, P., (1962) in Astronomie, Encyclopédie de la Pléiade, dir. Schatzman, E., Gallimard, Paris.*

orbits of visual binaries with known distances and luminosities, and if we assume a mass–luminosity relation, we can check the existence of such a relation and verify its shape.

The very accurate astrometric observations that are becoming available allow us to do better: it is possible in some cases to observe separately the motion of the components and to obtain in this way the position of their centre of mass, so that the mass of each of them can be determined. In such cases, we can speak of *astrometric binaries*.

5.1.2 Spectroscopic binaries

Many double stars are too close to each other to be separated in a telescope. With a spectrograph we observe in this case the superimposition of the spectra of the two components, and it is often possible to separate the spectral lines of each of them. Their wavelengths vary with time because of the Doppler–Fizeau effect corresponding to the orbital motion, which is such that one component comes closer when the other goes away (we ignore here the constant shift due to the global radial velocity of the system). The periodic variation of the radial velocity of the more massive component is smaller than that of the less massive one, and the ratio of the amplitudes is the inverse of the mass ratio, which can be determined if both series of lines are visible. The shape of these variations allow us to obtain the eccentricity of the orbits. However, it is generally impossible to know the inclination i of these orbits on the plane of the sky, so that only the projections of the semi major axes, $a_1 \sin i$ and $a_2 \sin i$, can be determined. From them we deduce the quantities $M_1 \sin^3 i$ and $M_2 \sin^3 i$, which are called the mass functions, using the third Kepler's law (Equation 5.2). We get in this way:

$$M_2 \sin^3 i = (a_1 \sin i)^3 (1 + M_1/M_2)^2/P^2, \qquad (5.3)$$

and a similar expression for $M_1 \sin^3 i$. Here, the masses are expressed in solar masses, the semi major axis in astronomical units and the period in years. The mass functions therefore are lower limits of the true masses.

There is a variant of this method for binary pulsars or double stars of which one component is a pulsar. The pulses they emit are so regular that the delays or advances resulting from their orbital motion can be easily measured, hence also the variations of their radial velocity.

Sometimes, only the radial velocity of one of the components can be measured because the other is too weak or even invisible if it is a black hole. If astrophysical considerations allow us to determine the mass M_1 of the visible component, for example the mass–luminosity relation if it is on the main sequence, we can obtain from Equation (5.3) a lower limit for mass M_2 of the invisible component: a lower limit because of the projection factor $\sin^3 i$. In this way, it was possible to show that the invisible components of some binary systems are significantly more massive that the maximum mass of a neutron star, i.e. 2 to 3 $M_\odot$ (see Section 2.6), thus they must be black holes.

Note that this method is also used to detect exoplanets which rotate around nearby stars and to obtain a lower limit for their masses (nearly 900 are known at the time of writing).

5.1.3 Eclipsing binaries

It is sometimes possible to observe one of the components of a double star go by the other. A case known for a long time is Algol (β Persei). Figure 5.3 shows the light curve of Algol, which has a period of 2.867 days, and its interpretation.

In the case of Algol, as it is for all eclipsing binaries, we are guaranteed that the line of sight is close to the plane of the orbit, so that $i \approx 90°$

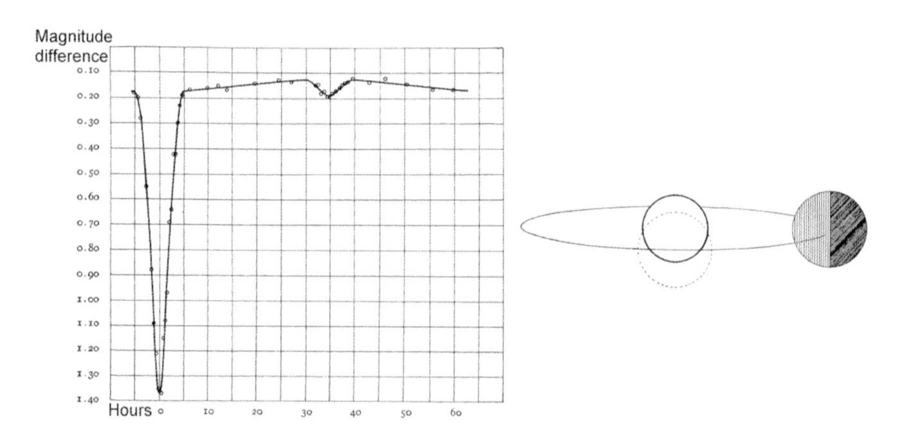

Figure 5.3. The light curve (periodical intensity variations) of Algol (β Persei) and its interpretation by Stebbins.[1] The deeper minimum occurs when the larger component (radius $R = 3.48\ R_\odot$), which is much less luminous than the other one, goes by the latter (the dashed circle in the image at the right). The secondary minimum occurs when it goes behind. The flux between the two minima is slowly variable because the hemisphere of the less luminous component which faces the more luminous one is brighter than the other hemisphere, as represented schematically. *From Stebbins, J. (1910) Astrophysical Journal 32, 185, with permission of the AAS.*

[1] This interpretation is somewhat obsolete: Algol is an interacting system, and its components are actually distorted by the interaction.

and sin $i \approx 1$. If the variations of the radial velocity of the components can be observed, e.g. if the binary is spectroscopic, the masses of both stars can be determined to good accuracy. The linear dimensions and eccentricity of the orbit are known from the variations of radial velocity and period, so that it is not necessary to measure the distance of the system independently. For Algol, as an example, recent determinations give $M = 3.17\ M_{\odot}$ for the more luminous component (spectral type B8V) and $M = 0.70\ M_{\odot}$ for the other one (K0IV). There is actually a third component with a much longer period of revolution (680 days) on a distant circular orbit, which produces no eclipse and does not affect the behaviour of the close binary much.

All the masses used to build the mass–luminosity relation of Figure 2.4 were obtained from eclipsing binaries.

The transit of exoplanets in front of their central stars has been observed in many cases either from the ground or from the satellites CoRoT (France and European Space Agency) and KEPLER (NASA). In such cases, we know that the inclination of the orbit is close to $i \approx 90°$ and the mass of the planet can be determined.

5.2 Mass transfer in double stars

Some binary stars have components so close to each other that their gravitational interaction is very important and gives rise to all sorts of interesting phenomena. We will now discuss these cases.

5.2.1 Equipotential surfaces around a close binary

Let us consider two stars A and B with respective masses M_1 and M_2, that we assimilate for the moment to two massive points (the same results would be obtained for spherical stars). They rotate around their centre of mass S, whose position between the two stars is given by:

$$a_1 M_1 = a_2 M_2 \text{, with } a = a_1 + a_2 \text{ the distance between the stars (Figure 5.4).} \quad (5.4)$$

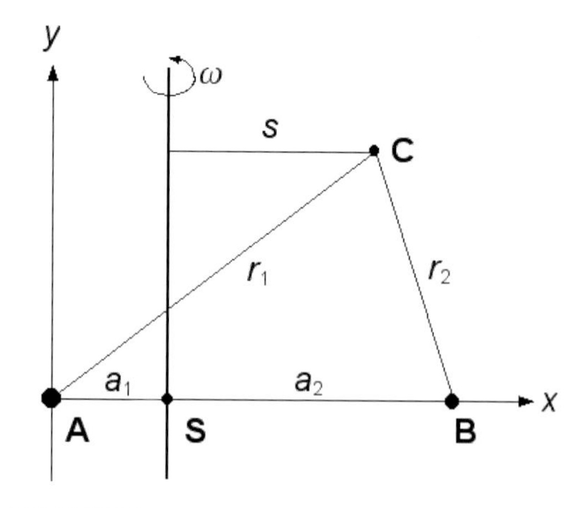

Figure 5.4. Scheme for the demonstration of Equation 5.6.

Consider a point C with coordinates x, y, z in a rectangular system of axes rotating with the system, taking A as its origin so that the two stars are on the Ox axis. The distances of C to the stars are respectively r_1 and r_2, so that $r_1^2 = x^2 + y^2 + z^2$ and $r_2^2 = (a - x)^2 + y^2 + z^2$; its distance to the rotation axis, which is directed towards z and goes through the mass centre S, is such that $s^2 = (a_1 - x)^2 + y^2$. The effective potential $\Omega(x, y, z)$ at point C is:

$$\Omega(x, y, z) = -\,\mathrm{G}M_1/r_1 - \mathrm{G}M_2/r_2 - \Omega_s, \tag{5.5}$$

where Ω_s is the potential of the centrifugal force, $\Omega_s = 1/2\ \omega^2 s^2$. The orbital frequency ω is given by the third Kepler's law (Equation 5.2): $\omega^2 = \mathrm{G}(M_1 + M_2)/a^3$. One derives from these relations the expression of the potential:

$$\Omega/\mathrm{G} = M_1/r_1 + M_2/r_2 + (M_1 + M_2)s^2/2a^3. \tag{5.6}$$

The corresponding equipotential curves in a cross section perpendicular to the rotation axis and containing the stars are drawn in Figure 5.5. One of these curves, which has the form of an 8, delineates the Roche lobes, from the name of the French astronomer Edouard Roche (1820–1883). The crossing point of this equipotential curve is Lagrange's L_1 point.

Figure 5.6 shows different possible cases according to the more or less complete filling of the Roche lobes by one or the other component.

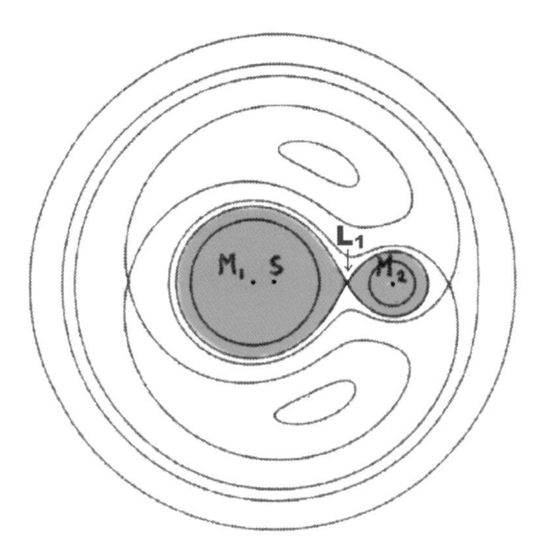

Figure 5.5. The equipotential surfaces of a double star. The two components are M_1 and M_2, M_1 being the more massive. S is the centre of mass. The Roche lobes are in grey, and the Lagrange point L_1 is indicated. *From de Boer & Seggewiss (2008).*

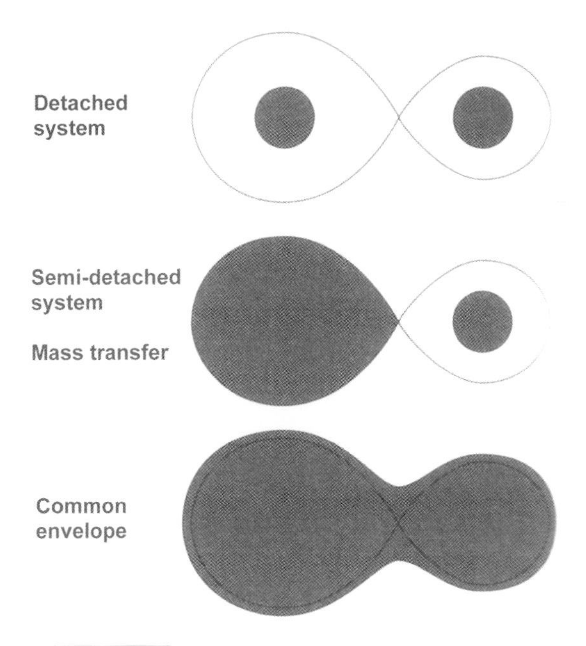

Figure 5.6. The different cases of interacting binaries.

When one of the components, which is always the more massive at the beginning of the evolution, becomes a red giant and fills its Roche lobe, it begins to pour its matter onto the other component. The volume of the Roche lobe around mass M_1 is approximately that of a sphere of radius R_L so that:

$$R_L/a \approx 0.38 + 0.2 \log(M_1/M_2), \tag{5.7}$$

and a similar expression for the M_2 lobe by interchanging M_1 and M_2. This equation allows us to calculate at what time in the evolution this phenomenon begins. Figure 5.7 illustrates as an example the evolution of a close double star whose components have masses 2 and 1 $M_\odot$, respectively.

The evolution can differ greatly according to the initial masses of the components. If the components are very massive, it can also be affected by mass loss. We cannot in this book discuss all the cases, and we will only examine the main effects of the infall of matter on one of the components.

5.2.2 Matter accretion on the compact component of a close binary

When the matter that escaped from the component which fills its Roche lobe falls on the other component, which is more compact, it possesses an angular momentum due to the revolution of the binary around its mass centre. This momentum prevents the matter from falling directly on the compact component, and forces it to form a rotating ring around it. The viscosity of the gas spreads out this ring which eventually forms a disk. These steps are illustrated on Figure 5.8.

Then the progressive outward transfer of the angular momentum of the disk is such that the matter in its inner parts eventually fall onto the compact component. During these processes, gravitational energy is liberated and transformed into heat and magnetic energy, the latter being itself transformed into other forms of energy. We call the *efficiency* of the process the ratio of the liberated gravitational energy to the energy of mass of the matter falling on the star. The gravitational energy produced by the matter, which had at the beginning a relatively small kinetic energy that can be neglected, is equal to the escape energy which would be needed to

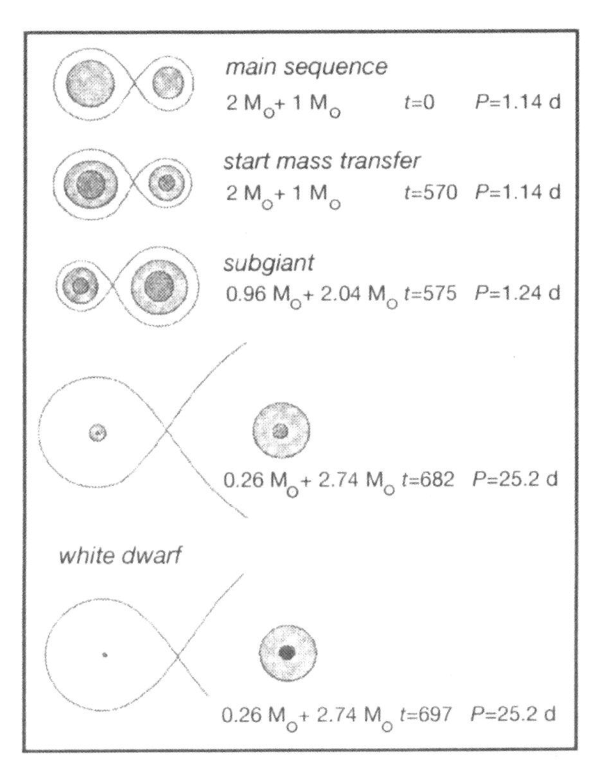

Figure 5.7. Evolution of a close double star. For each step schematized at the left, we give the mass of the components. The time *t* is in million years, while the period of revolution is in days. The Roche lobes are drawn. From top to bottom: 1. Initial stage: the two components, formed at the same time, are on the zero age main sequence; 2. The primary component, which is initially the more massive, is now a red giant with a helium core, and its envelope just fills its Roche lobe: the mass transfer to the other component starts; 3. The component which was initially less massive has gained much mass from the other component. Both components are sub-giants, i.e. red giants of low luminosity since their volume is limited to their Roche lobe. The mass transfer decreases and the period begins to increase; 4. The primary component is now reduced to its helium nucleus of 0.26 $M_\odot$, partially degenerate, surrounded by a large and tenuous hydrogen envelope which fills the Roche lobe. The mass transfer stopped and the period increased significantly; 5. The envelope falls on the helium core, which contracts and becomes a white dwarf. The secondary component becomes a red giant whose envelope will eventually fill, after 130 more million years, its enormous Roche lobe. A new episode of mass transfer will take place to the white dwarf (not shown): a nova or a bright source of soft X rays will result. The final stage will be a pair of white dwarfs, with all the rest of the matter expelled. *From de Boer & Seggewiss (2008).*

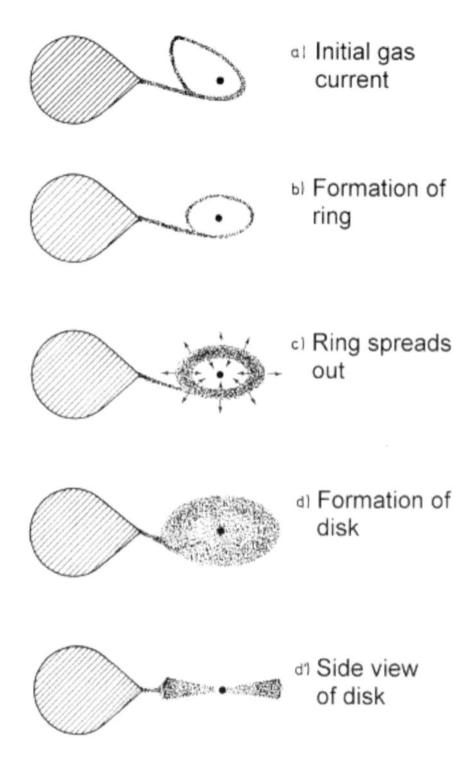

Figure 5.8. The different steps of the formation of an accretion disk in a close binary. *From Verbunt, F. (1982) Space Science Reviews, 32, 379.*

take this matter from the surface of the star and to put it at infinity. It is simply:

$$E = GMm/R, \tag{5.8}$$

where M and R are respectively the mass and the radius of the compact component and m the mass falling on its surface. For a white dwarf, this energy is of the order of 10^{-4} of the energy of mass mc^2, so that the efficiency is about 10^{-4}. As a comparison, the efficiency of the thermonuclear reactions which transforms hydrogen into helium is 7×10^{-3}: thus matter accretion on a white dwarf is much less efficient than its burning at its surface. We will see however that the two mechanisms are complementary, because the accreted matter experiences from time to time a thermonuclear explosion. For a neutron star, which is much more compact than a white dwarf, the efficiency is much larger, about 20%, and in this case thermonuclear reactions, when they exist, play a negligible role: when the accumulated matter explodes, this produces only a temporary energy excess which gives a burst of X-ray emission.

When matter falls on a black hole, the efficiency is in principle 100% because this matter disappears entirely. However this is only in principle, because the energy must get out. The only useful energy is that which is dissipated in the disk, because when the matter falls into the black hole it carries all its gravitational energy with it. The important quantity is thus the inner radius of the disk. General relativity predicts the existence of a last stable orbit, inside which the matter can only plunge into the black hole. The radius of this orbit depends on the rotation of the black hole: the faster this rotation, the smaller the radius of this orbit, until it is equal to the Schwarzschild radius (that of the event horizon) $2GM/c^2$ when there is co-rotation of the matter with the black hole. The efficiency is then 42%. This is a maximum: the efficiency is only 6% for a non-rotating black hole.

If the matter falls with a flux dm/dt, and if we suppose that all the gravitational energy is transformed into thermal energy and then into radiation, the luminosity of the object is:

$$L = GM/R \, dm/dt. \tag{5.9}$$

We may assume that the captured material is distributed more or less uniformly over the surface of the star, which can be the case if it is a white dwarf of a neutron star. The radiation is thus close to that of a blackbody at a temperature T such that:

$$L = 4\pi R^2 \sigma T^4, \tag{5.10}$$

where σ is the Stefan–Boltzmann constant. For typical rates of mass transfer of 10^{-8} to 10^{-10} $M_\odot$ per year, Equation (5.9) predicts luminosities from 10^{33} to 10^{35} erg/s for a white dwarf and from 10^{36} to 10^{38} erg/s for a neutron star or a black hole. The corresponding temperatures given by Equation (5.10) are of the order of 7×10^4 K for a white dwarf, corresponding to a photon energy of 6 eV, and of 10^7 K (1 keV photons) for a neutron star. Accreting white dwarfs thus emit a lot of visible and ultraviolet light, while neutron stars and black holes emit mainly in the X-ray range. However, we observe in some cases an emission of gamma rays of higher energy. This can come either from the channeling of matter by strong magnetic fields onto a reduced area of the surface of the star, where the matter would reach a much higher temperature, or from non-thermal processes in the case of neutron stars, as we will see later. Note finally that the infall of matter often occurs in a very irregular way, yielding strong luminosity fluctuations.

5.2.3 Close binaries and gravitational waves

As we have seen, the mass transfer from one component to the other affects the parameters of the system because the masses of the components are modified. Let us take as an example the case of a close binary in which the donating star is the less massive, as in some cases of Figure 5.7. In such a system, the centre of mass is close to the compact receiving object. As long as the mass of the donating star decreases, this star moves progressively off the centre of mass and its Roche lobe grows as a consequence. This should stop the mass transfer, especially because the radius of the donor decreases due to its mass loss. However, observation shows that in such binaries the X radiation which characterises the mass transfer persists when theory predicts that the transfer should cease. Why is this?

The answer was found in 1967 by Bohdan Paczyński (1940–2007): in such systems, the total angular momentum is not preserved. General Relativity predicts that all binary systems emit gravitational waves which carry with them rotational energy, hence also angular momentum. As a result, the components rotate faster and faster around each other and their distance decreases. Consequently, the Roche lobe of the donor does not grow, so that this component can go on pouring matter onto the other.

This idea was spectacularly confirmed when the decrease of the period of the binary pulsar[2] PSR B1913+16 was discovered, for which the Nobel

---

[2] A binary pulsar is a close double star in which at least one component is a pulsar. These objects are in general pairs of neutron stars.

prize in physics was granted to the Americans Joseph Taylor and Russel Hulse in 1993. Pulsars being extremely stable clocks, the monitoring of this binary pulsar over more than 30 years showed that the change in its orbital period followed exactly the prediction of General Relativity (Figure 5.9). Another result of their study is a very accurate determination of the mass of the two component neutron stars: 1.4398±0.0002 and 1.3886±0.0002 $M_{\odot}$ respectively. Many other binary pulsars are known, and several have allowed us to confirm the conclusions of the study of PSR B1913+16. There cannot be any doubt about the emission of gravitational waves by

Figure 5.9. The time decrease of the orbital period of the binary pulsar PSR B1913+16 observed for more than 30 years. The points are observational data, for which the error bar is smaller than the size of the symbol, and the parabola is the prediction of General Relativity. The ordinate gives the cumulative time differences between the observed time of passage at the periastron and that expected if there were no loss of angular momentum. *From Weisberg, J.M., et al. (2010), Astrophysical Journal 722, 1030, with permission of the AAS.*

close binary stars. Of course, it would be nice to detect them directly, and this is one of the reasons for the space project LISA of detection of gravitational waves.

The ultimate fate of close binaries is their fusion. This is inescapable as their distance can only decrease due to the energy lost through gravitational waves. This catastrophic phenomenon leads in its ultimate phase to a rapidly growing emission of accelerated periodic gravitational waves, which will hopefully be observed by ground-based gravitational detectors like VIRGO or LIGO.

5.3 Cataclysmic binaries, novae and Type Ia supernovae

We are now in a position to understand how the close binaries of the different types behave. The diversity of phenomena is so large that we cannot describe them all. Let us begin with the case where a white dwarf, a neutron star or a black hole receives matter from the other component which is a low-mass stars, and where the period of revolution is shorter than about ten hours. These systems are semi-detached (see Figure 5.6). We often call these particular double stars the *cataclysmic binaries*, because their emission is generally sporadic, or in any case very variable.

5.3.1 Dwarf novae and transient X-ray sources

In some of these cataclysmic binaries, the variations are due to thermal instabilities in the accretion disk which surrounds the compact component, instabilities which temporarily increase the accretion rate. The accretion disk is in thermal equilibrium if the heating through viscosity is balanced by the radiation losses. This is the case when the temperature is higher than about 10 000 K, because in this case the radiation cooling is efficient and increases with temperature, stabilising it. Thermal stability also occurs when the temperature is lower than about 5 000 K. At intermediate temperatures, which correspond to a partial ionisation of hydrogen by collisions, it can be shown that the disk is thermally unstable: any increase or decrease of the temperature propagates over the whole disk, which experiences relaxation oscillations between hot and cold configurations, as long as the rate of transfer of matter from the other component is such that the temperature is in the critical region. This gives rise to moderate, more or less regular eruptions. This is the case for *dwarf novae*, in which the compact component is a white dwarf. Their luminosity can increase by a factor 10 to 1 000 for a few months, to reach a value equal to 1 to 10 times that of the Sun.

If the compact object is a neutron star or a black hole, the object emits X rays during periods of activity that last a few months separated by quiescent periods which can be as long as tens of years: these are the *transient X-ray sources*. The sensitivity of present space X-ray telescopes is such that the emission of these objects can be detected even during quiescent episodes, during which some matter falls continuously on the compact object. Then, the luminosity in the visible is dominated by the donating star; its variations of luminosity and of radial velocity can be measured without being disturbed by the other component. It is thus possible to determine, as for all spectroscopic binaries, the total mass and that of the compact component. It is found that the mass of the latter is most often larger than 3 $M_{\odot}$, implying that it is a black hole rather than a neutron star. When this is the case, the X-ray luminosity during quiescent periods is particularly weak, as expected because almost all the gravitational energy of matter engulfed by a black hole is lost.

5.3.2 Novae

Classical novae are much more spectacular than dwarf novae. In a few days, a star which is most often less luminous than the Sun, and which had not been noticed before in general, reaches nearly 100 000 solar luminosities, then decreases regularly over a few months. Spectroscopy shows that matter has been ejected with a velocity of a few hundred to several thousand km/s. Novae are close binaries in which the compact component is a white dwarf. The hydrogen-rich matter from the other component accumulates on its surface and its density is such that it becomes degenerate. When its density and temperature are sufficient, hydrogen fusion starts. Because the reaction rates increase very fast with temperature, which rises with the production of energy, the reactions become explosive. Fusion is first by the proton–proton process, then by the CNO cycles. When the temperature reaches the Fermi temperature for which the degeneracy disappears, an expansion begins but the corresponding cooling is not sufficient to compensate for the energy production by the thermonuclear reactions, so that this expansion becomes explosive.

The mass of the ejected layer depends on the mass of the white dwarf, and varies between 10^{-4} and 10^{-5} $M_\odot$. It is lower if the stellar mass is higher, hence if its radius smaller (see Section 2.6 for a discussion of this apparent paradox). Because the accretion rate is 10^{-9} to 10^{-10} $M_\odot$ per year, we expect that all close binaries with a white dwarf should explode as novae every ten to hundred thousand years. Some novae are faster: these are the *recurrent novae*, whose mass is close to the Chandrasekhar limit, hence a particularly small ejected mass, and for which the accretion rate is high: they can reappear after a few tens of years.

The nucleosynthesis in the CNO cycles of novae is probably at the origin of several isotopes encountered in small quantities in the Galaxy: ^{13}C (only in part), ^{15}N and ^{17}O. A partial mixing of accreted matter with the matter of the white dwarf itself produces elements heavier than oxygen, e.g. ^{20}Ne. This nucleosynthesis probably stops in the region of sulphur, because the temperature is insufficient to go beyond this.

5.3.3 X-ray bursts

About half of the low-mass X binaries experience X-ray bursts of short duration — about 10 seconds — which occur at intervals of less than one hour. They can be explained in the same way as novae, although the compact object is now a neutron star instead of a white dwarf. As the pressure of the accreted matter is much higher than for a white dwarf, everything goes faster than for a nova: in a few days, the pressure and the temperature are so high that not only hydrogen fusion occurs, but also helium fusion. The temperature increases further, and the reactions accelerate, leading to an explosion. This explosion is recurrent because not all the material is ejected. The emission is in the X-rays (see the discussion in Section 5.2.2) and the luminosity of the burst is close to the Eddington limit for hydrogen-poor matter (Section 3.2).

Some bursts are much longer and repeat only every few years. These *super-bursts* are probably due to carbon explosive burning in deep regions under the accreted layer.

5.3.4 Type Ia supernovae

SN Ia are novae which turn bad: instead of the sole accreted layer of the white dwarf, the explosion concerns the whole star. It is certainly a thermonuclear explosion — this was proposed 40 years ago, but the details are poorly understood.

The standard scenario considers accretion of matter onto a white dwarf made essentially of ^{12}C and ^{16}O. The weight of the accreted matter compresses the white dwarf whose radius is reduced from 10 000 km to 3 000 km while its temperature increases. When the mass of the star reaches the Chandrasekhar limit, about 1.4 $M_\odot$, it cannot be supported by the electron pressure of the degenerate matter anymore and collapses, which produces a rapid increase of the central temperature. When this temperature reaches 4×10^8 K, carbon burning occurs, producing a lot of energy. However, because the matter is degenerate, the pressure does not increase and the matter does not expand, and hence cannot cool. The nuclear reaction becomes explosive: in less than a second, all the successive available fuels (^{12}C, ^{16}O, up to ^{28}Si) burn and produce heavier nuclei, in particular ^{56}Ni. This is so fast that the neutrinos emitted in the process do not have time to escape.

Then the flame of the combustion moves to the external layers. However, in spite of many years of investigation, we still do not know whether this motion is supersonic (*detonation*) or not (*deflagration*). In the first case, the external layers of the white dwarf do not know that the flame is coming and all the matter burns in succession to be transformed into ^{56}Ni and in other nuclei of the iron group (^{54}Fe, ^{58}Ni, etc.). In the other case, the external layers have time to be pushed outwards, their density decreases and the combustion stops before ^{56}Ni is reached, producing lighter nuclei of Ca, S or Si. The outermost layers do not experience combustion and are simply ejected. The spectral analysis of SN Ia shows the presence of these intermediate elements, but also of ^{56}Ni, so the truth is probably between the two cases: we think that the phenomenon starts with a deflagration at the centre that transforms itself into detonation nearer to the surface. But no satisfactory numerical model has yet been built for this.

A different scenario involves the collision followed by merging of two white dwarfs in a binary system, whose kinetic energies had been lost through the emission of gravitational waves. This has the advantage of accounting for the absence of hydrogen in the matter ejected by the supernova. However, energy loss via gravitational waves is a very slow process, and this scenario cannot account for the presence of SN Ia in very distant galaxies, which are necessarily very young. Worse, no explosion of the merged stars occurs in numerical models.

Whatever the scenario, all the matter of the object is dispersed into space with a kinetic energy of about 10^{51} ergs, which is essentially the thermonuclear energy produced by the explosion. By a curious coincidence, this energy is similar to that of the other, massive supernovae, while the mechanisms are completely different.

We do not understand any type of supernova well, but our ignorance is qualitatively different in the different cases. For thermonuclear SN Ia, we understand more or less how the star becomes a supernova, but the mechanism of the explosion is poorly understood. The opposite is true for the massive supernovae.

One may wonder why certain white dwarfs expel the accreted matter sporadically, while others keep it until their final explosion.[3] This might be due to a different accretion rate, which would be higher in SN Ia progenitors than in cataclysmic binaries and novae. It seems that the rotation of the white dwarf might also play an important role. These problems are presently the subjects of intense investigations. It would be very interesting to be able to observe the precursor of a SN Ia, like for the massive supernova SN 1987a in the Large Magellanic Cloud (see Section 4.2.4). The accretion rate being quite large in such a precursor, it should be a powerful X-ray source: a source of very soft (low-energy) X rays, because accretion is on a white dwarf. Such *super soft X-ray sources* are known, and we can only hope that one of them explodes.

---

[3] It might however be that some recurrent novae, whose mass is close to the Chandrasekhar limit, end up as SN Ia.

In spite of all these uncertainties, the SN Ia are used as standard candles in observational cosmology. Cosmologists are looking for very luminous celestial objects with identical luminosities, in order to obtain the luminosity/redshift relationship which is one of the elements that allows us to chose between different models of the Universe. SN Ia are good candidates for this, contrary to the other types of supernovae whose properties are too variable. It is however disquieting to see that cosmologists use objects whose properties are so poorly understood. Their optimism comes from the observations themselves: although the light curves differ from supernova to supernova, they fit into a universal curve after a correction which uses an empirical relationship between their maximum luminosity and their characteristic decay time (Figure 5.10).

5.4 Gamma binaries and microquasars

The close binaries described in the preceding section consisted of a low-mass star pouring its matter onto a compact object. There also exist close binaries in which the donor is a massive star. We saw that massive stars emit strong winds; in a close binary a part of the wind can captured by the compact object. Such binaries emit X rays. For most of them, the compact object is a neutron star which might be a X-ray pulsar, or a black hole like in the famous object Cygnus X-1. Recent observations showed that several such systems emit gamma rays: at least three cases are known, for which most of the energy is emitted as gamma rays of energy larger than several MeV. In these three objects, the compact star is a pulsar. The gamma emission comes from the region where the matter ejected by the pulsar with relativistic velocities interacts with the wind of the massive component.

The *microquasars* are different binaries. They are close double stars in which the accretion disk around the compact object loses a large amount of angular momentum by emitting two symmetrical jets along its rotation axis, like in protostars (Figure 5.11). This is a very reduced version of quasars: in quasars, the accretion disk surrounds the giant black hole located at the centre of a galaxy, with masses of millions to billions of solar masses.

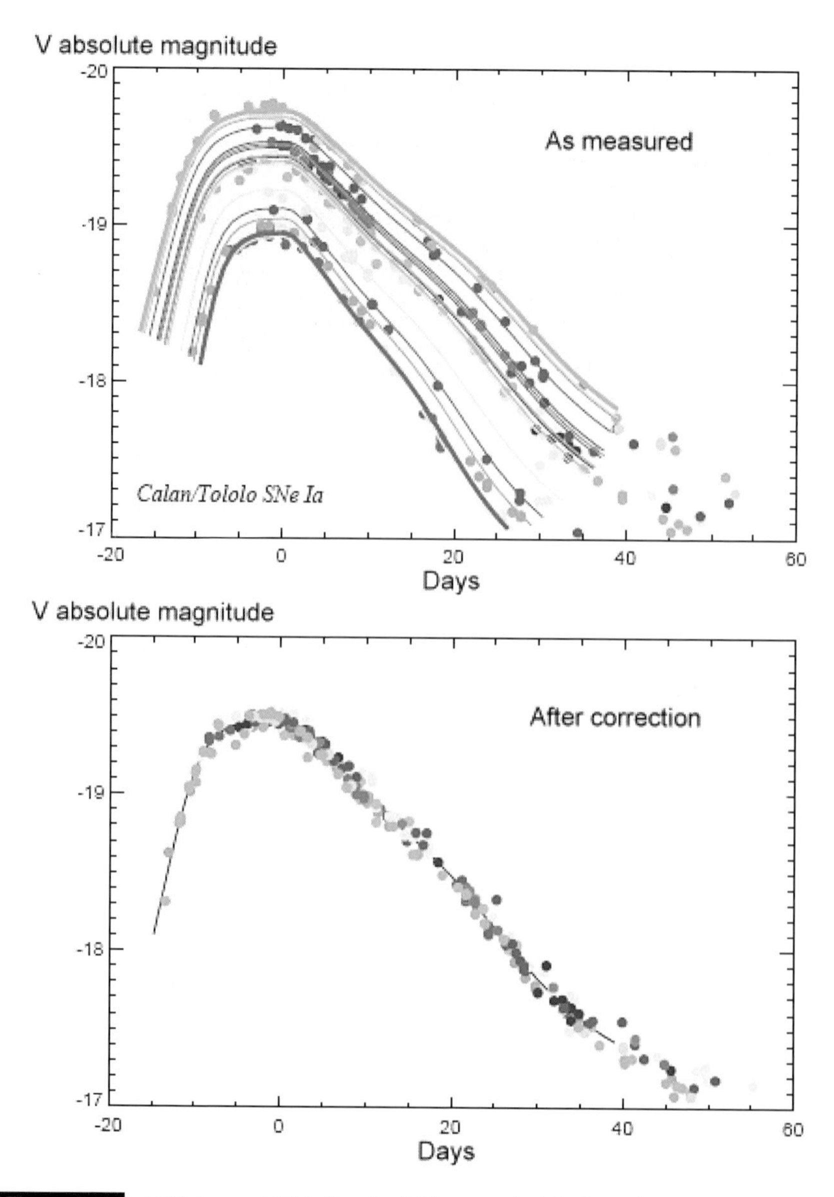

Figure 5.10. Light curves of SN Ia. Top, the light curves of relatively nearby supernovae reduced at the same distance: the V absolute magnitude is plotted as a function of time. Bottom, the same after an empirical correction described in the text. *From Lasota, J.-P., in Lequeux et al. (2009).*

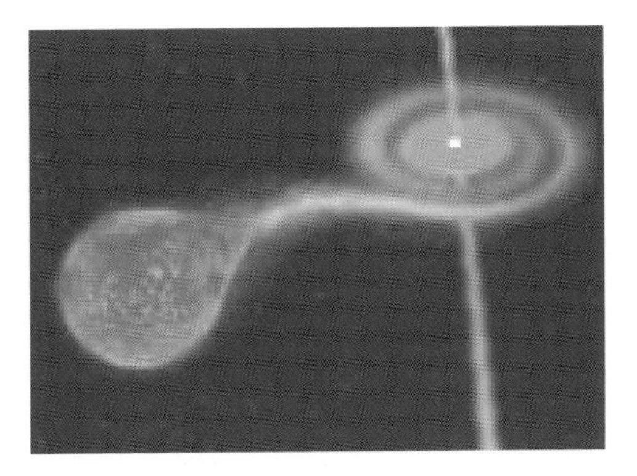

Figure 5.11. Diagram of a microquasar. The component to the left pours matter onto the compact component, forming an accretion disk whose angular momentum is transferred to two symmetrical jets.

5.5 Millisecond pulsars

Some pulsars emit radio pulses at an extraordinarily fast pace, which is their rotation period: their very focused emission is produced in two beams which rotate like those of a lighthouse. The fastest known pulsar makes 716 rotations per second, while most of them rotate at 300 turns/second. These *millisecond pulsars* belong to close binaries. They are pulsars whose rotation has been accelerated: the material which falls onto the neutron star did not lose all its angular momentum in the accretion disk, and this momentum accelerates its rotation. Indeed, such an acceleration has been observed in many X-ray binaries. The pulsar needs 10 to 100 million years to reach its fastest rotation speed. During the period of acceleration, it emits X rays and perhaps radio waves but the latter are absorbed by the matter being accreted, which is a plasma. When accretion stops, either because the star rotates too fast or because the companion does not send matter anymore, the plasma disappears, so that the radio waves are not absorbed anymore and can be observed. If the rotation frequency were to reach 3 000 turns/second, the pulsar would explode due to the centrifugal force. But no frequencies as high as this are observed: if the pulsar is not strictly symmetrical, it may be that the rotation is limited by an energy loss through emission of gravitational waves, which we hope to detect in the future.

Figure 5.12. The globular cluster 47 Tucanae in the Southern hemisphere, observed with one of the 8-m telescopes of the ESO VLT. At least 22 millisecond pulsars have been discovered in this cluster. © *ESO*.

Several hundred millisecond pulsars are known. They are very numerous in globular clusters like 47 Tucanae (Figure 5.12), in which the density of stars is so high that close binaries are frequently formed by gravitational capture. Apart from the periodical frequency variations due to their membership to binary systems, the millisecond pulsars are extremely stable clocks. Several observatories, including the Nançay radioastronomy station of the Paris Observatory, follow systematically the frequency of millisecond pulsars: this is of great interest for metrology and has several astrophysical applications. We hope to be able to detect small variations of their frequencies that would be due to the crossing of long-period gravitational waves.

6

Stars and the Evolution of Galaxies

6.1 Introduction

A galaxy like ours contains ten to a hundred billion stars, born from interstellar matter 13 billion years ago. The only first-generation stars that survive have masses smaller than 0.8 $M_{\odot}$. More massive stars have disappeared, having returned a part of their matter to the interstellar medium, while the rest forms a compact object: white dwarf, neutron star or black hole, whose lifetime is infinite or extremely long except in exceptional cases. Because higher-mass stars have a shorter lifetime than lower-mass ones, the surviving stars of a given generation have a higher minimum mass if they were formed more recently. In any case, an increasing fraction of the mass of a galaxy is contained in low-mass stars and compact objects. As a consequence, the mass fraction contained in the interstellar medium can only decrease with time. It may be, however, that fresh interstellar medium coming from outside is accreted by the galaxy. Conversely, the galaxy can lose interstellar medium by emitting a galactic wind. All this is summarised in Figure 6.1.

Let us now consider the evolution of the chemical composition of galaxies. In the beginning, there was only hydrogen and its isotope deuterium, helium 4 and 3, and lithium, which are the elements produced

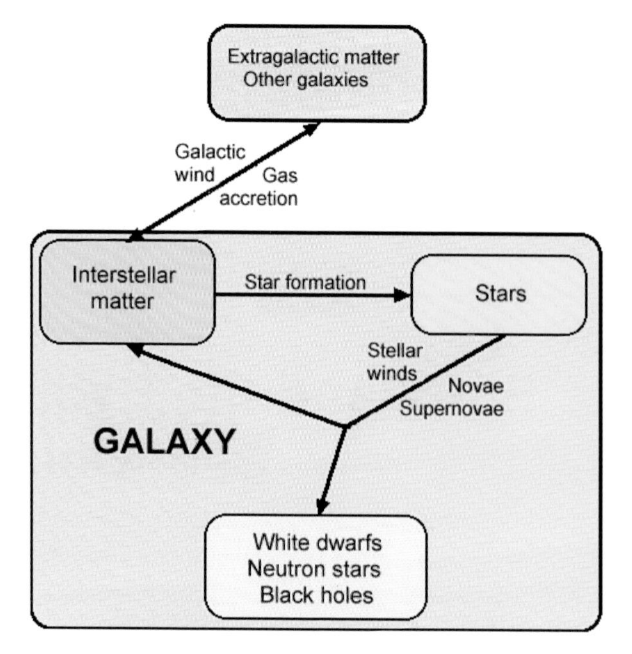

Figure 6.1. Diagram of the evolution of a galaxy.

just after the Big Bang. The stars synthetise heavier elements from these primordial ones, then eject them through their winds and when they die: we thus observe a progressive enrichment of the interstellar matter in heavy elements, which is reflected in the new stars born from it. However, there is a possibility that galaxies accrete fresh gas, perhaps primordial, which slows down their enrichment in heavy elements and can even decrease the abundances in the interstellar matter. Accretion is very important.

Modelling this chemical evolution is a complex affair, as many factors have to be taken into account:

- the lifetime of stars of different masses;
- the distribution of masses of stars at formation, which is the *initial mass function* (IMF);
- the *star formation rate* (SFR);
- the production (*yield*) of the different elements in stars;
- the ejection of these elements (stellar winds, novae, supernovae);

- their mixing with the interstellar gas;
- finally, the interaction of the galaxy with its surroundings (accretion or ejection of interstellar matter).

Almost all these factors may vary with time, which makes the problem even more complicated. In practice, we must simplify it by assuming that some of the parameters are constant (this is generally the case for the initial mass function without proof to the contrary), or by restricting their variations in a more or less arbitrary manner. We will now examine the various aspects of the problem, beginning with the production of the different elements.

6.2 The production of elements in stars

The foundations of stellar nucleosynthesis were laid in 1957 by Margaret and Geoffrey Burbidge, William Fowler and Fred Hoyle (paper B^2FH, see bibliography), a work which won Fowler the Nobel prize in physics in 1983. They recognised the principal processes (hydrogen and helium fusion, α, r, s and p processes, etc.) and identified the elements formed by these processes but did not specify the places where they occur, which were essentially unknown at the time. We described them in the previous chapters. Figure 6.2 presents a schematic curve of the abundances of elements in the Universe, with an indication of the processes which form them. Figure 6.3 is a periodic table where these processes are indicated by colours. Note that B^2FH did not know that a part of helium, deuterium and a part of lithium 7 were synthetised in the Big Bang, and ignored the origin of lithium 6, of the rest of the lithium 7, of beryllium and of boron, which we know today to come from breaking of interstellar nuclei by cosmic rays (*spallation* reactions). However, they noted that deuterium, lithium, beryllium and boron are very fragile and are most often destroyed in stars.

Let us recap which are the main sites for nucleosynthesis:
- hydrogen fusion: Big Bang, main sequence stars;
- spallation: interstellar medium;
- helium burning: red giants, AGB stars;
- explosive burning: novae, X-ray bursts, supernovae;

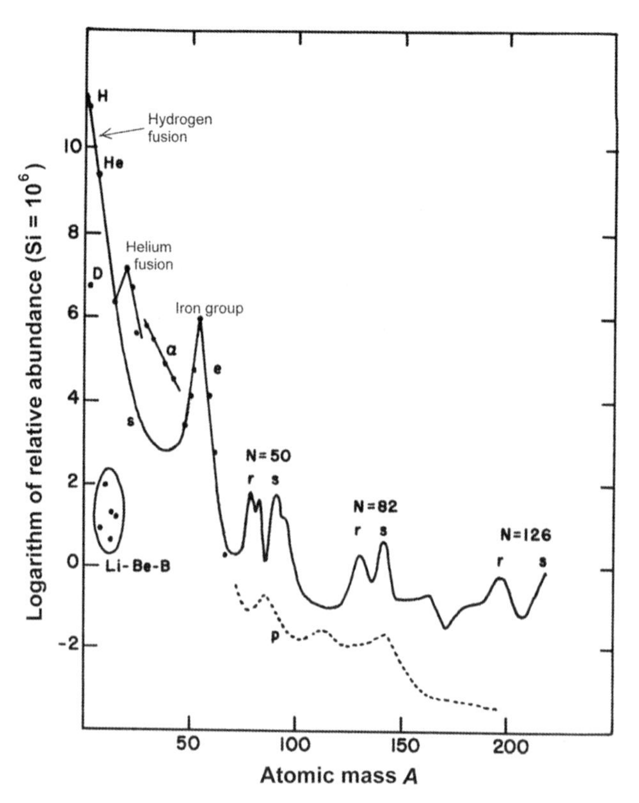

Figure 6.2. Schematic curve of abundances in the Universe with an indication of the nucleosynthesis mechanisms, from the B²FH paper (see reference in the bibliography). Though dating from 1957, this scheme is still valid. The atomic mass *A* is in abscissa. *N* is the number of neutrons in the nuclei. The formation processes are indicated by the letters α, *e*, *r*, *s* and *p* respectively. Note the abundance peak of the iron group, due to the high stability of the corresponding nuclei, and the peaks for α elements with $A = 16, 20, 24 \ldots 40$. There are also peaks for $A = 80$ and 90, 130 and 138, and 194 and 208, corresponding to nuclei containing the magic numbers of neutrons $N = 50$, 82 and 126; the s nuclei are richer in neutrons than the *r* nuclei.

- nuclosynthesis beyond iron, *s* process: AGB stars;
- nuclosynthesis beyond iron, *r* process: Ia supernovae;
- nuclosynthesis beyond iron, *p* process:?

We have now almost all that is needed to understand the enrichment of the interstellar medium by stars. Almost, because we have not yet discussed how stellar nucleosynthesis depends on the initial chemical

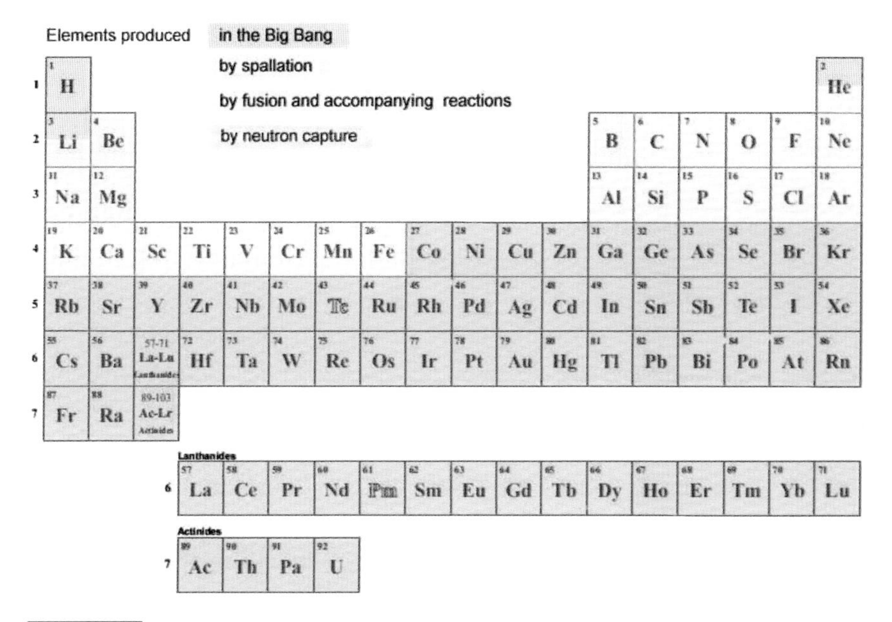

Figure 6.3. A periodic table showing the origin of the different elements. The *s* and *r* elements formed by neutron capture are not distinguished here. The artificial elements, which are all unstable, are not shown.

composition of stars. For example, stars of the first generation (which astronomers call *Population III* stars) contained only hydrogen and helium, with no heavy elements. None of these stars have ever been observed, because they were all very massive and hence had short lifetimes. However, the Galaxy contains very old stars which are very poor in heavy elements (the stars of *Population II*, in contrast with the younger *Population I* stars), and which corresponds to the second generation after Population III. Their chemical composition gives information about nucleosynthesis and enrichment of the interstellar medium in the first phases of Galactic evolution.

The first stars formed ^{4}He by the p–p process, then ^{12}C by the 3α process, allowing the CNO cycle to work and to synthetise more helium. Their evolution was so rapid that slow neutron capture was negligible, so that s elements were not formed. Only α elements were synthetised, and these stars contained only ^{4}He, ^{12}C, ^{16}O, ^{20}Ne, ^{24}Mg and ^{28}Si at the end of their evolution. There were also small quantities of ^{56}Ni resulting from the fusion of ^{28}Si. These Population III stars exploded as supernovae, synthetising r elements and rejecting them along with others into the interstellar medium. Indeed, the most deficient Population II stars of the second generation contain r elements formed in this way, no s elements, and some lithium 7 made in the Big Bang that they have not destroyed because their mass is too small.

It is impossible to know the initial mass function (IMF) of Population II stars because the most massive ones have disappeared. This is also true for the older generations. One has to assume that it was the same as the IMF for stars which are forming at present, which fortunately looks similar everywhere. Then it becomes possible to calculate what they reject into the interstellar medium, taking into account the fact that their mass loss becomes larger while their metallicity increases from generation to generation. Figure 6.4 shows the contribution of the stars with different masses to the enrichment of the interstellar medium for two different initial metallicities. The contribution of novae and of SN Ia is neglected here, because it is probably rather small with respect to that of ordinary stars, except for iron as we will see later. Integrating over the IMF, we get the total enrichment per stellar generation.

Apart for the uncertainty surrounding the IMF, there is another unknown in the problem, even if we admit that the models which were used to build Figure 6.4 are entirely correct: we do not know if, and above which mass, the matter of the most massive stars is completely engulfed in a black hole at the time of their explosion as supernovae. Any matter falling into a black hole is lost for the enrichment of the interstellar medium, and this can affect the production of heavy elements greatly, especially at low metallicities.

6.3 From stars to interstellar medium

The material ejected by stars into the interstellar medium can be in different forms. Stars of low and intermediate mass eject matter during the red giant and AGB phases as a relatively cold wind, and some elements can form molecules then dust grains: these grains are made of silicates or of carbonaceous products according to the chemical composition of the expelled envelope. Later, they will be covered with ices in molecular

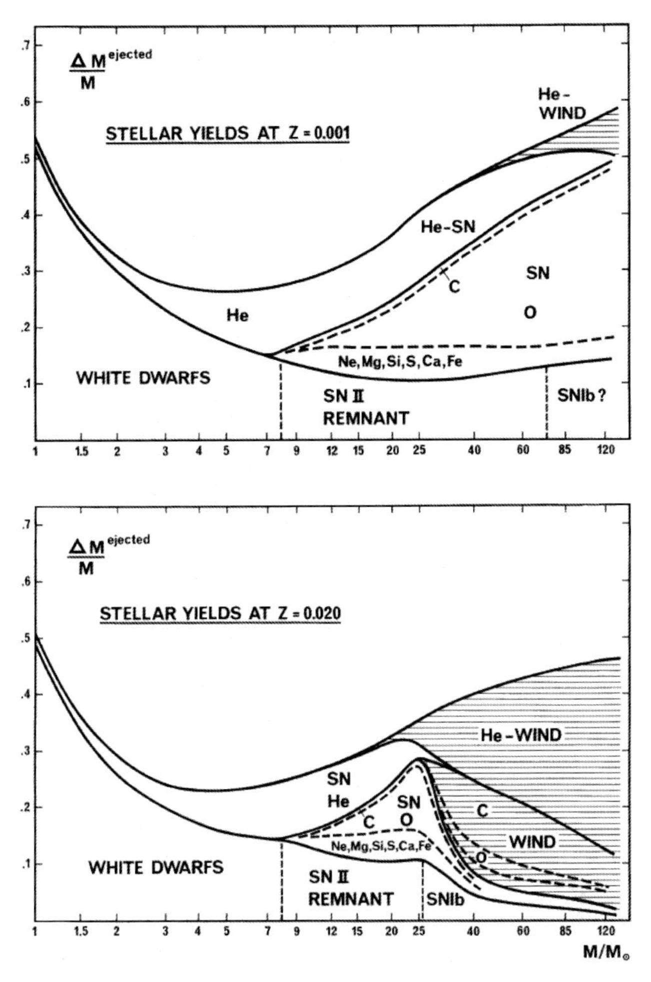

Figure 6.4. Production of heavy elements in the interstellar medium by stars of different masses with initial metallicities $Z = 0.001$ (top) and $Z = 0.020$ (bottom); the latter is approximately the metallicity in the present local interstellar medium. The initial mass is in abscissa, and the ordinate gives the mass fraction from the centre to the outside of the star. The fraction of mass blocked in the compact remnant after the death of the star is given by the lower curve. The ejected fraction of the initial mass is above this curve, and the part which enriches the interstellar medium in helium and heavy elements is between these two curves. Above the upper curve, the ejected matter has kept its initial composition and does not intervene in the enrichment. We notice that massive stars contribute to the enrichment mainly through their winds at high metallicities, but not at low metallicities. For the largest masses, the stars could disappear entirely in a black hole when they explode as supernovae (SN Ib?), in which case the production of heavy elements would be zero at this stage. As can be seen on the figure, this effect is more important at low metallicities. *From Maeder, A. (1992) Astronomy & Astrophysics 264, 105, with permission of ESO.*

clouds. Dust grains can also be destroyed by the ultraviolet radiation of hot stars, and new grains can condense in the interstellar medium. What happens to the dust is of no importance for chemical evolution, because it condenses together with the gas when forming new stars.

It is clear that the matter is expelled by the giants with a kinetic energy which is too small to allow it to escape the gravitational field of a galaxy. It is not necessarily the case for the ejecta of massive stars. Their winds are hot and have a considerable velocity, several thousand km/s. This is also true for the matter ejected in supernova explosions, which is extremely hot, with a temperature of 10^5 to 10^7 K: its thermal emission can be observed as X-rays, together with emission lines of very ionised elements which give information on the temperature and chemical composition. These ejecta interact with the surrounding medium to form supernova remnants, but a part remains intact in interstellar space where it occupies a large fraction of the volume. Frequently, many stars are born simultaneously in a given region of space (Figure 6.5), and the most massive

Figure 6.5. The cluster of young, massive stars NGC 3603. This image in false colour has been obtained in the visible and in the infrared with one of the 8-m telescopes of the ESO VLT. The distance of the cluster is 7 kpc, and the dimensions of the image are 11×11 pc. © *ESO.*

explode as supernovae after a few million years. The collective effect of winds and explosion forms a gigantic bubble filled with very hot gas (Figure 6.6). This bubble may break the colder gas of the galactic disk and form a chimney through which its content spreads into the galactic halo. The very hot gas cools slowly and condenses as neutral clouds which fall back into the galactic disk, sometimes far from its place of origin: this is what astronomers call the *galactic fountain*. Through these steps, the heavy elements synthetised in supernovae or contained in stellar winds find their way after several million years to an interstellar medium able to form new stars.

When the bubble is really very big, a part of the gas it contains succeeds in overcoming the gravitational potential of the galaxy, and can escape into intergalactic space, carrying with it elements newly synthetised by massive stars, and also by less massive ones because some of the gas they have enriched might have mixed with the hot gas.

Figure 6.6. The bubble N 70 in the Large Magellanic Cloud, imaged with one of the 8-m telescopes of the ESO VLT. Its diameter is 95 pc. It results from the explosion of numerous massive stars belonging to a cluster similar to that of Figure 6.5. Some stars not yet exploded are visible at the centre. © *ESO*.

Figure 6.7. The galaxy M 82, seen edge on with the Hubble Space Telescope. It is the seat of a very intense star formation in its central regions. Winds and explosions of massive stars result in the expulsion of a large quantity of gas, which is visible in red because it emits the Hα line of hydrogen. © *Hubble Space Telescope Heritage.*

Spectacular examples of these galactic winds are shown in Figures 6.7 and 6.8.

Galaxies can also accrete gas from outside. Observations show that this was a very important phenomenon for high-redshift galaxies, which are young and not very evolved. We have increasing evidence that accretion is still active today, for example in the capture of small galaxies by bigger ones.

6.4 The evolution of the chemical composition of galaxies

6.4.1 Basic equations

Let us write first the equations of mass conservation in a galaxy, or in a part of a galaxy, with M = total mass (excluding possible dark matter), M_g = mass of interstellar gas (including dust, assumed to be well mixed

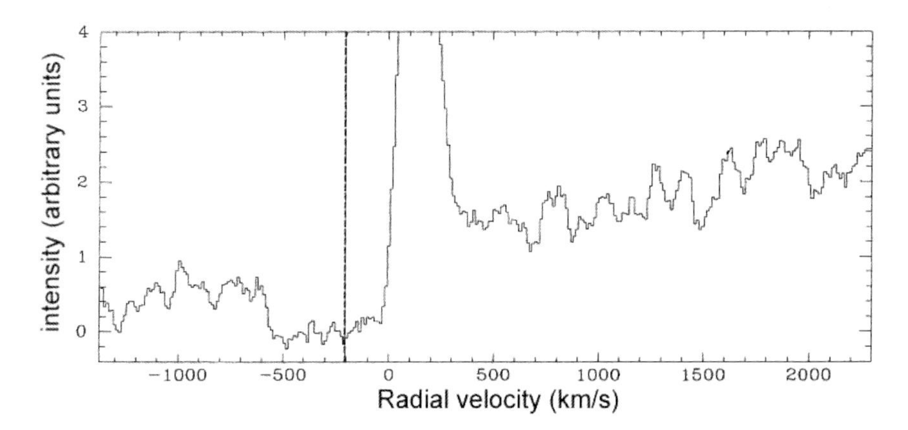

Figure 6.8. The profile of the Lyman α line of hydrogen of the blue compact galaxy Haro 2. Like M 82 (Figure 6.7), this galaxy is the seat of an enormous burst of star formation. The radial velocity with respect to the average velocity of the galaxy is given in abscissa, derived from the wavelength by the Doppler–Fizeau relation. The line has a P Cygni profile (see Figure 3.10), characteristic for mass loss: gas is expelled from the galaxy with a relative velocity of –200 km/s (vertical dashed line), and absorbs the continuum and the line emitted deeper in the galaxy. The velocity of –200 km/s (vertical dashed line) is that of the centre of the absorption corrected for the emission profile. *From Lequeux, J. et al. (1995) Astronomy & Astrophysics 301, 18, with permission of ESO.*

with the gas), $M_\star$ = mass of stars, f = rate of accretion of gas (+ dust) and e = rate of ejection of gas. ψ is the rate of star formation and E the rate of ejection of matter by stars. All thee quantities are in general functions of time. We can write:

$$M = M_g + M_\star, \quad (6.1)$$

$$dM/dt = f - e, \quad (6.2)$$

$$dM_\star/dt = \psi - E, \quad (6.3)$$

and

$$dM_g/dt = -\psi + E + f - e. \quad (6.4)$$

At time t, those stars of mass m which were formed at time $t - \tau_m$ die, τ_m being their lifetime.[1] Consequently, the death rate of stars at time t is equal to their formation rate at time $t - \tau_m$, i.e. $\psi(t - \tau_m)\varphi(m)$. Here, we

[1] Contrary to the notation in the preceding chapters, we call m the mass of a star instead of M, M being reserved to the total mass of an ensemble of stars or gas.

separate the part ψ (SFR) which depends on time and the IMF $\varphi(m)$, which we suppose to be independent on time. We normalise the IMF so that $\int m\varphi(m)dm = 1\ M_{\odot}$, the integral being over all the stellar masses of a generation. As to the SFR ψ, it can be taken as constant, or function of the mass of gas M_g, or function of any other parameter.

We often use Salpeter's IMF, which is sufficiently well described for masses larger than that of the Sun by the form:

$$m\varphi(m) = mdn(m)/dm = 0.17\ m^{-1.35}, \tag{6.5}$$

with the normalisation above; $n(m)$ is the number of stars with masses between m and $m + dm$.

We might alternatively use Scalo's IMF, which is perhaps more realistic:

$$\begin{aligned} m\varphi(m) &= 0.83 && \text{for } m \quad 0.5\ M_{\odot}, \\ &= 0.26\ m^{-1.7} && \text{for } m > 0.5\ M_{\odot}. \end{aligned} \tag{6.6}$$

A star of mass m ejects at its death $m - w_m$, w_m being the mass of the compact remnant. Here, we include the non-enriched mass into the ejected matter. In order to obtain the total amount of ejected matter $E(t)$, we have to integrate the mass ejected by all stars whose mass is larger than $m_{min}(t)$, the smallest mass of the stars which have disappeared since the formation of the galaxy at $t = 0$ (or, equivalently, the largest mass of the stars that survive from this epoch).

The ejection rate E at time t is therefore:

$$E(t) = \int (m - w_m)\psi(t - \tau_m)\varphi(m)dm, \tag{6.7}$$

the integral being taken from $m_{min}(t)$ to infinity. The mass of the compact remnant is approximately given by:

$$\begin{aligned} w_m &= 0.11m + 0.45\ M_{\odot}\ (m \leq 8\ M_{\odot}) \\ w_m &= 1.5\ M_{\odot}\ (m > 8\ M_{\odot}). \end{aligned} \tag{6.8}$$

Let us write now the equation for the element abundances in the interstellar medium. It is similar to (6.4) but applied to the abundance Z (remember that Z is expressed as a fraction of the mass of gas):

$$d(ZM_g)/dt = -Z\psi + E_Z + Z_f f - Ze, \qquad (6.9)$$

The production rate of this element is:

$$E_Z(t) = \int [(m - w_m)Z(t - \tau_m) + mp_Z(m)]\psi(t - \tau_m)\varphi(m)dm. \qquad (6.10)$$

The term between square brackets include the ejected fraction of element Z which was inside the star at its formation, $(m - w_m)Z(t - \tau_m)$, and the mass of this element added by the nucleosynthesis in this star, $mp_Z(m)$. Note that certain elements like deuterium, lithium, beryllium and boron are partly or totally destroyed in stars so that the latter term in negative for them. Note also that Equations (6.7) and (6.8) assume that the element Z ejected by the star is rapidly mixed with the interstellar medium.

The mass ejected per unit mass of stars of a given generation, i.e. by all stars born together at a given time, is:

$$R = \int (m - w_m)\varphi(m)dm, \qquad (6.11)$$

But be careful! The stars do not immediately eject mass into the interstellar medium. If we neglect the small fraction of synthetised elements that the stellar winds might carry with them, the restitution of elements occurs at the end of the life of the star, so that at a given time t the integral has to be taken only over masses larger than $m_{min}(t - \tau)$, τ being the time when the considered generation was formed.

Another interesting quantity is the *yield* y_Z, which is the ratio of the mass of the element Z (or all elements) newly synthetised to the mass which remains in stars and compact objects:

$$y_Z(t) = 1/(1-R) \int mp_Z(m)\varphi(m)dm, \qquad (6.12)$$

the integral being also only over masses larger than $m_{min}(t - \tau)$.

Case	R	y_{He}	y_Z	y_O	y_C
$Z = 0.001$					
A	0.45	0.052	0.030	0.019	0.0022
B	0.42	0.044	0.010	0.0035	0.0012
C	0.41	0.042	0.0079	0.0023	0.0010
D	0.40	0.040	0.0054	0.0012	0.00076
E	0.38	0.034	0.0018	0.00019	0.00028
$Z = 0.020$					
A	0.47	0.055	0.024	0.0071	0.0094
B	0.47	0.055	0.024	0.0071	0.0094
C	0.46	0.054	0.021	0.0048	0.0091
D	0.46	0.052	0.017	0.0035	0.0087
E	0.44	0.047	0.012	0.0019	0.0077

Table 6.1. Mass fraction R ejected into the interstellar medium, and yields y_{He}, y_Z, y_O and y_C of helium, of all elements heavier than helium, of oxygen and of carbon respectively, for the Scalo IMF. Z is the initial metallicity of the stars. Case A is that for which no massive star is entirely engulfed in a black hole when it explodes. Cases B, C, D and E correspond to the cases for which the stars are entirely engulfed in a black hole when their initial masses are larger than 27.5, 22.5, 17.5 and 11.6 $M_\odot$ respectively. *From Maeder, A. (1992 and 1993) Astronomy & Astrophysics 264, 105 and 268, 833.*

Table 6.1 gives numerical values for R and for the yield of helium, oxygen, carbon and the sum of all elements heavier than helium (y_{He}, y_O, y_C and y_Z respectively), for the Scalo IMF, *after all the stars which contribute to the enrichment have finished their evolution* ($m_{min} \approx 1\ M_\odot$ in practice).

6.4.2 Primary and secondary elements

Figure 6.4 and Table 6.1 assume that the detailed chemical composition of the matter from which the stars were formed is similar to that in the solar vicinity (see Table 1.1). However, this composition does not necessarily correspond to reality. While the production of He and O is practically independent of the initial composition, other elements are formed principally, and in some cases only, from pre-existing heavy elements. The first ones are the *primary elements*, and the other ones are the *secondary elements*. ^{14}N is generally considered as a secondary element: the CNO cycles and secondary reactions synthetise ^{14}N, and also ^{12}C, ^{13}C, ^{22}Ne, etc., from ^{16}O (see Section 2.4.2 and Box 4.1). ^{14}N is also a by-product of ^{12}C burning.

The yield of a secondary element produced by a star is thus expected to be proportional to the initial abundance of its parent primary element, and this proportionality should be reflected in the chemical evolution of a galaxy. Observations show that this is approximately the case for nitrogen with respect to oxygen, at least when the interstellar gas has a relatively high metallicity (Figure 6.9). However, at low metallicities, the abundance of nitrogen appears to be independent of that of oxygen: this shows that there is also a production of primary nitrogen, probably by stars of large and intermediate masses. This production dominates at low metallicities because there is little oxygen, and becomes negligible with respect to the secondary production when the abundance of oxygen is large. The dispersion of N/O at a given value of O/H is real and is not well explained.

This example shows that we must be very careful when using yields to study the chemical evolution of galaxies. If oxygen seems to be a rather pure primary element, many elements are partially primary and partially secondary as can be seen in diagrams similar to that of Figure 6.9. This is the case for carbon, while neon, sulphur and argon appear to be

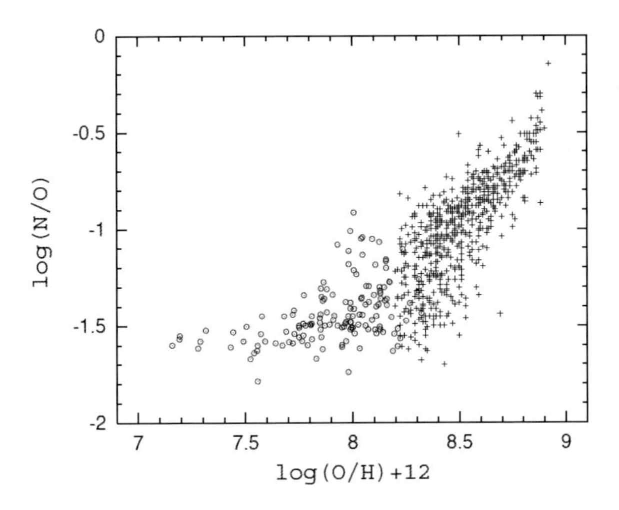

Figure 6.9. Relation between the abundance ratio of nitrogen to oxygen and the abundance of oxygen, for HII regions in irregular galaxies (circles) and in spiral galaxies (crosses). *From Pilyugin, L.S. et al. (2003) Astronomy & Astrophysics 397, 487, with permission of ESO.*

primaries. Clearly we do not yet understand all the subtleties of stellar nucleosynthesis.

Iron itself does not behave like oxygen, but this does not imply that it is not primary, because the production sites of these two elements are rather different. Oxygen, like other α elements (^{12}C, ^{20}Ne, ^{24}Mg, ^{28}Si and ^{32}S), is formed in stars more massive than 8 $M_\odot$, whose lifetimes are shorter than 20 million years. Hence oxygen appears early in the evolution of a galaxy. Iron is produced in supernovae, but the low-mass SN Ia dominate its production. Remember that the progenitors of SN Ia are white dwarfs which result from the evolution of relatively low-mass stars: SN Ia thus appear late in the galactic evolution. As a result, the ratio between the abundance of the α elements and that of iron decreases continuously with time in a galaxy: the α/Fe ratio offers an important diagnostic of the history of star formation. One can estimate empirically the relative contribution of massive and of low-mass supernovae to the production of iron: in the oldest halo stars of the Galaxy, whose heavy elements were produced exclusively by massive stars (SN II, Ib or Ic), the ratio O/Fe is 4 times larger than in the Sun, while it is only 0.8 times the solar ratio in very young stars which have been preferentially enriched into iron by the SN Ia.

6.4.3 The instant recycling approximation

It is easy to solve numerically the equations of chemical evolution given in Section 6.4.1. However the results obtained in this way are not always intuitive. In order to obtain more easily understandable results, we often use the *approximation of instant recycling*, which assumes that the synthetised elements are injected without delay into the interstellar medium. We write now $\psi(t - \tau_m) \approx \psi(t)$. R and y_Z do not depend on time anymore, and represent respectively the fraction of their total mass ejected by stars of all masses after they have evolved, and the yield of these stars in the element Z. The approximation is valid for elements created by massive stars whose lifetime is short, but is less good when we consider stars of smaller masses.

In this approximation, Equations (6.7) and (6.10) become respectively:

$$E(t) = R\psi(t) \tag{6.13}$$

and

$$E_Z(t) = RZ(t)\,\psi(t) + (1-R)y_Z(t)\,\psi(t). \tag{6.14}$$

Inserting the latter equation into (6.9) we obtain:

$$d(ZM_g)/dt = (1-R)[-Z + y_Z(t)]\,\psi(t) + Z_f f - Ze. \tag{6.15}$$

Equations (6.3) and (6.4) become:

$$dM_\star/dt = (1-R)\,\psi(\mathrm{t}), \tag{6.16}$$

$$dM_g/dt = -(1-R)\psi(\mathrm{t}) + f - e, \tag{6.17}$$

then, combining (6.15) and (6.17):

$$M_g dZ/dt = (1-R)y_Z(t)\,\psi(t) + (Z_f - Z)\,f - Ze. \tag{6.18}$$

Equations (6.16), (6.17) and (6.18) are the three fundamental equations of the approximation of instant recycling. If we ignore the exchanges with the external world ($f = 0$ and $e = 0$), there is a further simplification: this is the *closed box model*. In this case we can divide Equation (6.17) by Equation (6.18) after suppression of the f and e terms, and we obtain:

$$(1/M_g)dM_g/dZ = -\,1/y_Z, \tag{6.19}$$

a differential equation whose solution is, assuming y_Z constant:

$$\ln(M_g/M) = -Z/y_Z. \tag{6.20}$$

We thus obtained in this very simple model a relation between the amount of gas in a galaxy and the abundance of heavy elements: the smaller the gas metallicity, the more gas there is in a galaxy, an intuitive result indeed.

The equations of galactic evolution allow us to simulate the time evolution of the abundances in the gas or in stars at their formation, and to simulate for example the age–metallicity relation or the distribution of abundances in stars of different masses at a given time. In the instant recycling model with evolution in a closed box, it suffices to assume a relation between the star formation rate and the mass of gas to obtain the age–metallicity relation. Posing:

$$\psi(t) = \omega(t) M_g, \tag{6.21}$$

the abundance grows with time like:

$$Z(t) \propto \int_0^t \omega(t')dt'. \tag{6.22}$$

Assuming for example ω constant (formation rate proportional to the mass of gas), Z grows linearly with time. This is approximately the case for stars in the solar vicinity, but there is a large dispersion in the abundances for stars with the same age, so that we are forced to admit a mixing of the stellar populations, an idea amply confirmed by recent observations.

Under the same approximations, the distribution of stellar abundances should not depend on the time history of star formation. It can be obtained by writing Equation (6.20) as:

$$M_g/M = 1 - M_\star/M = \exp(-Z/y_Z),$$
$$\text{or } M_\star/M = 1 - \exp(-Z/y_Z), \tag{6.23}$$

Then differentiating and multiplying by Z, $M_\star$ being a function of Z:

$$dM_\star(Z)/d(\ln Z) = MZ/y_Z \exp(-Z/y_Z). \tag{6.24}$$

For stars which have similar masses, for example the red giants of the central regions (the bulge) of our Galaxy, this equation gives the distribution of the number of stars as a function of their metallicity. This distribution has a maximum for $Z = y_Z$. For bulge stars, it agrees with observations for $y_Z = 0.015$ (Figure 6.10). Given the fact that metallicity increases with time, this value of the yield is compatible with the models of Figure 6.4 and Table 6.1, provided that the stars with an initial mass larger than 25 $M_\odot$ disappear entirely into a black hole after their

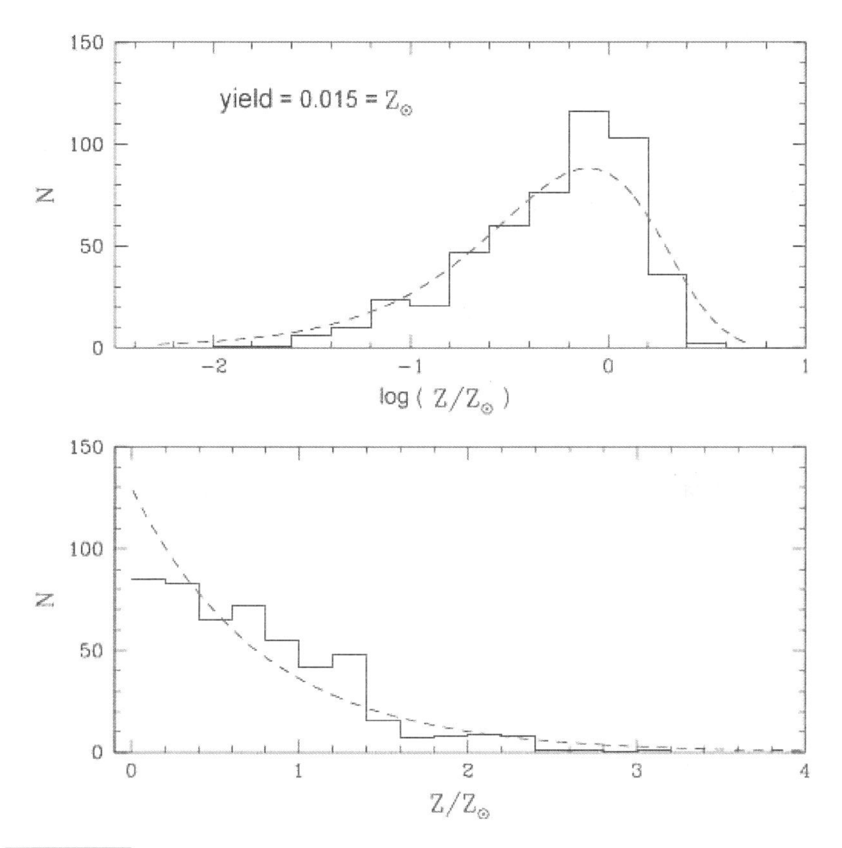

Figure 6.10. Distribution of metallicity for the red giants of the galactic bulge. Top, the number N of stars per logarithmic interval of metallicity. The curve is the prediction of the instant recycling model in closed box (Equation 6.24). Bottom, the number of stars per linear interval of metallicity, and the prediction of the model (Equation 6.23). The agreement between model and observations is good. *From Zoccali, M. et al. (2003) Astronomy & Astrophysics 399, 931, with permission of ESO.*

explosion as a supernova. However, one can also imagine that the efficiency is larger but that the gas of the bulge experienced mass loss during the life of the Galaxy.

6.4.4 Some examples of application

We stressed the uncertainties with our knowledge of stellar nucleosynthesis, in particular the fact that it is not easy to know if an element is primary, secondary or both. To add to this problem, we can only observe the abundances of a small number of elements:

- in gaseous nebulae, thus in the present interstellar medium, we can measure the abundance of He, C, N, O, Ne, S, Ar and sometimes Fe. When we speak of "interstellar abundance", it is generally that of oxygen, which is considered as the primary element *par excellence*;
- in stars, except for the Sun and bright stars, we can essentially measure C, O (not in all stars), Mg, Ca and Fe. When we speak of "stellar abundance", it is generally that of iron, which does not behave like oxygen as we have seen.

Given the uncertainties in the abundances and in other parameters, it is often sufficient to use the evolution model with instant recycling. Although it might be misleading in some cases, we may use it in order to have an idea of the metallicity that a galaxy should have, given its gas content, and vice versa. If the predictions of Equation (6.20) are poor, chances are the evolution is not closed box.

Let us consider first the observed relation between the oxygen abundance and M_g/M, which are both known for many galaxies. The oxygen abundance has been measured to a reasonable accuracy in many galaxies, but the mass of gas is not so well known given the uncertainty on the masses of molecular clouds, and the total mass as measured dynamically can contain a poorly determined amount of dark matter. Figure 6.11 shows the result. We remark that if an oxygen yield $y_O \approx 0.0027$ accounts rather well for this relation for spiral galaxies in the instant recycling model, it seems too high for irregular galaxies, the more so if they are less evolved and richer in gas. This can be explained either by a systematic effect in the production of oxygen by stars as a function of metallicity or, more probably, by a loss of enriched interstellar matter by little-evolved irregular galaxies: they have a mass, and hence a gravitational potential, that is particularly small so that they lose their gas more easily. Strong galactic winds are indeed observed for some of these galaxies (see Figures 6.7 and 6.8). But even the value $y_O \approx 0.0027$ seems too small compared to the data of Table 6.1, suggesting that spiral galaxies can also lose gas.

The effect is visible for the two irregular galaxies satellite of the Milky Way, the Magellanic Clouds. For the Large Cloud, we observe

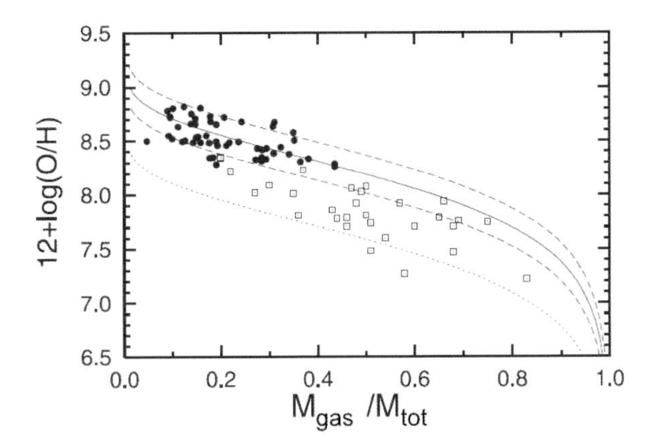

Figure 6.11. Oxygen abundance O/H as a function of the gas fraction M_g/M for spiral galaxies (black circles) and irregular galaxies (squares). The full curve is the prediction of the instant recycling model in closed box, with $y_O = 0.0027$; the dashed curves on both sides correspond to values of this yield respectively 1.5 times larger and smaller. The dotted curve is for a yield four times smaller. *From Pilyugin, L.S. et al. (2004) Astronomy & Astrophysics 425, 849, with permission of ESO.*

$Z_O \approx 0.0034$ and $M_g/M \approx 0.13$, requiring $y_O \approx 0.0017$ if there is no gas exchange with the exterior. For the Small Cloud, we have $Z_O \approx 0.00115$ and $M_g/M \approx 0.43$, hence $y_Z \approx 0.0014$. These yields are significantly smaller than 0.0027, in agreement with what we see in Figure 6.11.

A similar problem occurs for the metallicity distribution of stars in our Galaxy. Let us consider first the halo stars, which are metal-poor. Their metallicity distribution agrees rather well with Equations (6.23) or (6.24) provided that there is little gas left as observed, but with the condition that the oxygen yield is very small, about 0.0004. A possible explanation is a loss of gas during the evolution. However, it seems that most of these stars have been captured by our Galaxy from low-metallicity dwarf galaxies, which only displaces the problem in general.

For the stars in the galactic disk, which have a larger metallicity, one may wonder why low-metallicity stars are so rare compared to the predictions of Equations (6.23) or (6.24). The answer is not straightforward, because this is a case where simple models of galactic evolution fail. The stellar population we observe close to the Sun is a mixture of stars coming from different galactocentric radii: they migrate radially due to the lack of axial symmetry of the galactic gravitational potential stemming from the spiral arms and the central bar. We are not yet in a position to understand completely what happens, in spite of a large quantity of recent observational work. It is likely that the astrometric satellite GAIA will provide answers to many irritating questions on this topic.

6.5 Evolution of colour and spectrum of galaxies

Another aspect of the evolution of galaxies is the time variation of their luminosity, colour and spectrum. We observe that relatively unevolved galaxies, which still contain a lot of gas, are bluer than the others, and that their spectrum shows the characteristics of young stars. Let us discuss first the evolution of the colour of galaxies.

6.5.1 The colour of galaxies and its evolution

Since it became possible to measure the integrated flux of galaxies through coloured filters in the middle of the XXth century, it has been remarked that galaxies lie on a well-defined sequence in a colour–colour diagram. Figure 6.12 shows this sequence in the $U-B$, $B-V$ diagram, but one would obtain similar diagrams by plotting any pair of different colours. In this sequence, the colour of galaxies is blue at the top left ($B-V \approx 0.3$, $U-B \approx -0.4$) and red at the bottom right ($B-V \approx 1.0$, $U-B \approx 0.7$). We will see that the blue galaxies are much less evolved than the red ones, a property that we can check in looking at their M_g/M ratios.

In order to study quantitatively this evolution, Richard Larson and Beatrice Tinsley (1941–1981) calculated the colours, and also the V-band luminosity, of a stellar population born with Salpeter's initial mass function. In one of their simulations, all the stars were born at the same time in a burst of star formation, and then evolve with time without further star formation. In another case, the stars form continuously with a constant rate. Table 6.2 shows the results. Of course, the colours become redder more rapidly and the luminosity decreases faster in the first case.

It is remarkable that the colours of "normal" galaxies follow the law of Table 6.2 for the case of a constant stellar formation: it is the corresponding track which is plotted in Figure 6.12. One might be tempted to

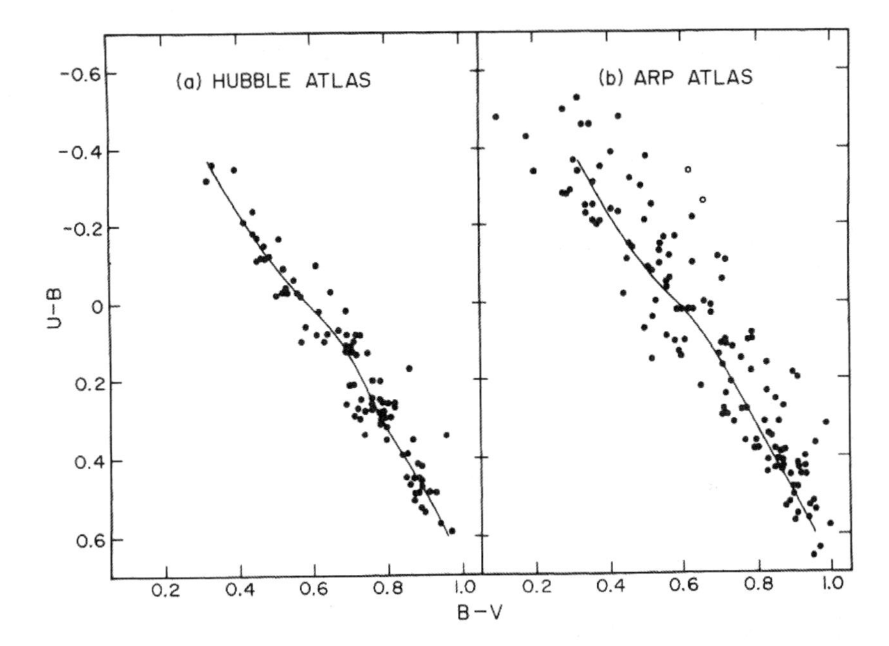

Figure 6.12. Representative points of galaxies in the colour–colour diagram *U–B*, *B–V*. Left, "normal" galaxies of the Hubble Atlas; right, galaxies of the Arp Atlas, which show important morphological particularities due to their interaction with other galaxies and sometimes also merging with another galaxy. We notice that the representative points are much more dispersed for the latter galaxies. *From Larson, R.B. & Tinsley, B.M. (1978) Astrophysical Journal 219, 46, with permission of AAS.*

conclude that the bluer galaxies started to form stars only recently, while the redder ones have formed stars for a long time. This would be a premature conclusion. Indeed, calculations for star formation rates decreasing regularly with time give almost identical results. *The position of the representative point of a galaxy on the curve of Figure 6.12 depends in practice only on the ratio of the star formation rate during the last 10^8 years to the mass of all stars formed before*. Therefore, it is impossible to derive the past history of star formation from the colours alone: one can only determine its integral over time. Fortunately, for galaxies closer than about 20 megaparsecs, we can study the stellar population directly and obtain further information. The result is astonishing: even in the bluest galaxies for which we might think that formed stars only recently, a small amount of old stars is invariably found.

Age (10^9 years)	Burst of star formation			Continuous star formation		
	B–V	*U–B*	L_V/M ($L_{V\odot}/M_\odot$)	*B–V*	*U–B*	L_V/M ($L_{V\odot}/M_\odot$)
0.01	–0.20	–1.00	45	–0.20	–1.00	45
0.02	–0.08	–0.79	28	–0.15	–0.93	36
0.05	+0.19	–0.37	18	–0.03	–0.78	28
0.1	+0.26	–0.19	9.1	+0.04	–0.67	19
0.2	+0.37	–0.02	6.3	+0.11	–0.56	14
0.5	+0.51	+0.17	3.0	+0.21	–0.43	7.7
1.0	+0.62	+0.22	1.4	+0.27	–0.36	5.0
2.0	+0.73	+0.29	1.1	+0.34	–0.29	3.2
5.0	+0.86	+0.42	0.53	+0.44	–0.20	1.7
10	+0.94	+0.56	0.26	+0.50	–0.14	1.0
20	+1.02	+0.74	0.12	+0.56	–0.09	0.59

Table 6.2. *U–B*, *B–V* colours and V luminosity/mass ratio for an ensemble of stars formed with Salpeter's IMF. In the first case, the stars are all born together in a burst at age 0 and evolve without further star formation, and in the second case stars are formed a a constant pace since age 0. *From Larson, R.B., & Tinsley, B.M. (1978) Astrophysical Journal 219, 46.*

How can we interpret the dispersion of the representative points of "anomalous" galaxies in the right part of Figure 6.12? We can assume that in these cases a burst of star formation is occurring, or occurred recently, and contributes strongly to the luminosity and colours. Figure 6.13 illustrates this hypothesis. A comparison of this figure with the right part of Figure 6.12 confirms this idea: we can always explain the colours of these galaxies by superimposing on a burst of star formation of different strength and age onto a normal galaxy. This explanation dates from 1978, but has been fully confirmed by further observations. We know at present that gravitational interactions between galaxies produce gigantic concentrations of interstellar matter where bursts of star formation occur. Figure 6.14 shows a spectacular example.

All these conclusions, which have been established for the U, B and V bands, are valid for any colour from the ultraviolet to the near infrared.

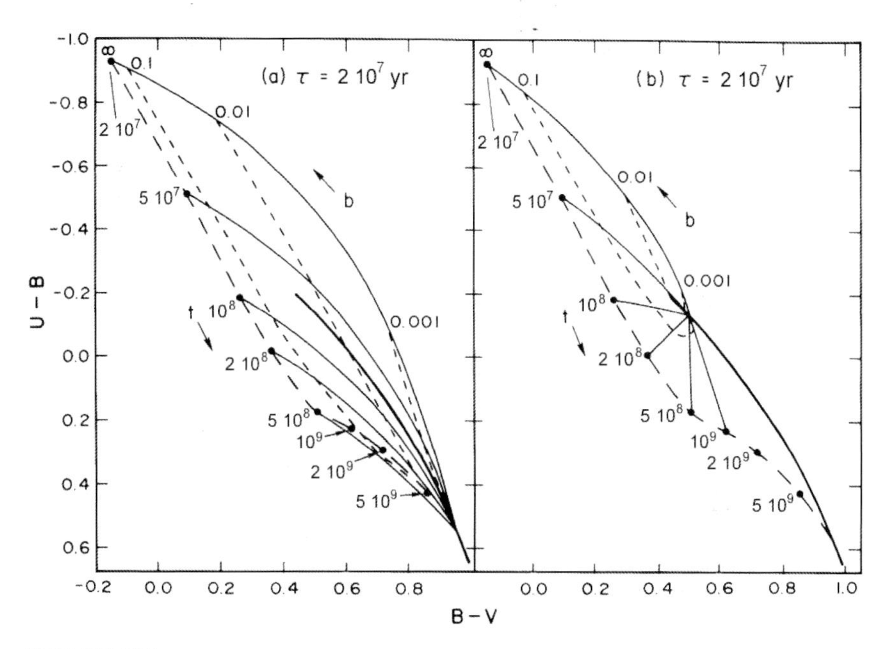

Figure 6.13. The effect of a burst of star formation on the colour of a galaxy. Left, the emission of a burst is superimposed onto the light of a red galaxy with $B–V = 0.95$ and $U–B = 0.54$. The upper curve indicates the resulting colours for very young bursts aged 2×10^7 years, with different intensities: the numbers on the curve give the ratio b of the mass of stars formed in the burst to the total mass of stars formed previously in the galaxy. For a infinite value of b, the colours are those of a pure burst, see Table 6.1. The lower curves represent the colours obtained for bursts of different intensities as a function of the age t of the burst, which is indicated for each curve. The dashed curves show the time evolution of the colours for bursts of intensity $b = 0.001$, 0.01, 0.1 and ∞. The bold curve is the locus of the colours of the normal galaxies (see Figure 6.12). Right, the same diagram for bursts superimposed on a blue galaxy with $B–V = 0.50$ and $U–B = –0.14$. Notice that the colour can be redder if a strong burst occurred a long time ago. *From Larson, R.B. & Tinsley, B.M. (1978) Astrophysical Journal 219, 46, with permission of the AAS.*

6.5.2 The spectral evolution of galaxies

The power of present computers allows us to synthetise the spectrum of a stellar population and follow its time evolution. It is also possible to synthetise the spectrum of galaxies with different star formation histories and to compare it with observations. This is in a sense an extension of the work on the colours of galaxies which was discussed in the preceding section.

Figure 6.14. A burst of star formation in the interacting galaxies NGC 4038–4039. The central regions of the two galaxies are the white diffuse spots. The burst, which started a few million years ago, created many young, blue stars. The red regions are gaseous nebulae ionised by the most massive of these stars; their colour is due to the Hα line of hydrogen. Dust filaments absorb the background light. © *Hubble Space Telescope Heritage.*

Figure 6.15 shows the time evolution of the spectrum of stars born simultaneously with a Salpeter initial mass function, and the evolution of the spectrum of a galaxy which has formed stars at a constant pace since the origin. This figure is the equivalent of Table 6.1, which showed the evolution of the colours and luminosity for the same cases.

Of course, it is possible to superimpose a burst of star formation onto a continuous star formation that is decreasing at a constant speed with time. Here are some results of a comparison of such simulations with actual spectra of galaxies:

- The spectra of elliptical galaxies are generally well represented by an ancient burst of star formation, including their ultraviolet radiation which was often considered as mysterious in the past. These galaxies stopped forming stars a long time ago, in agreement with

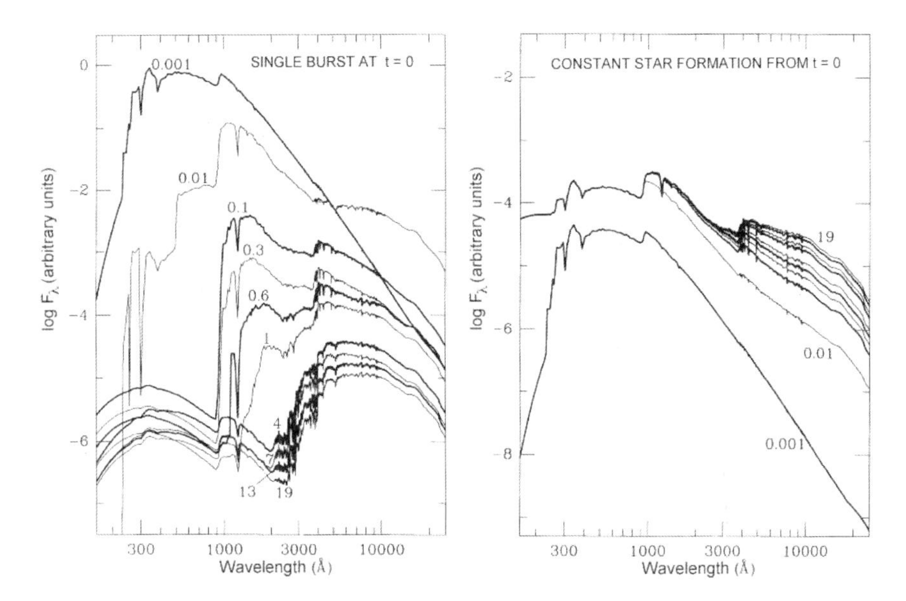

Figure 6.15. Time evolution of the spectrum of stars born with a Salpeter initial mass function (log–log scales). Left, evolution of the integrated spectrum of stars born simultaneously as a function of their age in billions of years, indicated for each curve. At the beginning the spectrum is dominated by hot, massive stars which radiate essentially in the far ultraviolet. Then they progressively disappear, and stars of increasingly lower masses dominate, which emit in the near ultraviolet, the visible and the infrared. There remains however a faint radiation in the very far ultraviolet due to the nuclei of planetary nebulae. Right, evolution of the integrated spectrum of a galaxy where constant star formation began at an age indicated near some of the curves, in billions of years. The calculations go until 19 billion years, although we know that the age of the Universe is only about 14 billion years. *From Bruzual, G.A., & Charlot, S. (1993) Astrophysical Journal 405, 538, with permission of the AAS.*

the fact that they contain little gas. However, some elliptical galaxies continue to form stars, probably because they have captured gas from outside.

- The spectra of irregular galaxies are well represented by a young burst. This does not mean that no star formation was present before: the best way to check this is to detect directly the old stars by deep imaging. Every time this has been done, old stars have been found, as we remarked earlier. No really young galaxy has ever been discovered.

- The spectra of spiral galaxies can always be represented by continuous star formation. But their variation with time is not well constrained by model fitting to the observed spectra.

Finally, one finds in these comparisons the same results, with almost the same uncertainties, as in a comparison of colours with evolutionary models. Nothing can replace the direct imaging of stellar populations, which is for the moment limited to relatively nearby galaxies, say closer than 20 megaparsecs. To go further, we will have to wait for the giant telescopes presently in construction.

Bibliography

Arnett, D.A. (1996) *Supernovae and Nucleosynthesis*, Princeton University Press.

De Boer, K.S. & Seggewiss, W. (2008) *Stars and Stellar Evolution*, EDP Sciences, les Ulis: *cited in the captions of figures as de Boer & Seggewiss (2008).*

Clayton, D.D. (1984) *Principles of Stellar Evolution and Nucleosynthesis*, McGraw Hill & University of Chicago Press.

Cox, J.P. (1980) *Theory of stellar pulsations*, Princeton University Press.

Hansen, C.J. & Kawaler, S.D. (1999) *Stellar Interiors, Physical Principles, Structure and Evolution*, Springer, Berlin & Heidelberg.

Kippenhahn, R. & Weigert, A. (1990) *Stellar Structure and Evolution*, Springer, Berlin & Heidelberg.

Lequeux, J., with the collaboration of Falgarone, E. & Ryter, C. (2005) *The Interstellar Medium*, Springer, Berlin & Heidelberg.

Lequeux, J. (coord. et auteur), Acker, A., Bertout, C., Lasota, J.-P., Prantzos, N. & Zahn, J.-P. (2009) *Étoiles et matière interstellaire*, Ellipses, Paris: *cited in the captions of figures as Lequeux et al. (2009).*

Maeder, A. (2009) *Physics, Formation and Evolution of Rotating Stars*, Springer, Berlin & Heidelberg (a much more general study than the title indicates): *cited in the captions of figures as Maeder (2009).*

Mihalas, D. (1970) *Stellar Atmospheres*, W.H. Freeman and Co.

Pagel, B.E.J. (2009) *Nucleosynthesis and chemical evolution of galaxies*, Cambridge University Press.

Stahler, J.W. & Palla, F. (2004) *The formation of Stars*, Wiley Weinheim.

Weiss, A., Hillebrandt, H.-C., Thomas, A.-C. & Ritter, H. (2004) *Cox & Giuli's Principles of Stellar Structure, Extended Second Edition*, Cambridge Scientific Publication.

A few interesting references to complete the preceding books

Bruzual, G.A. & Charlot, S. (1993) Spectral evolution of stellar populations using isochrone synthesis, *Astrophysical Journal*, 405, 538–553; freely accessible via http://cdsads.u-strasbg.fr/abs/1993ApJ...405..538B

Burbidge, E.M., Burbidge, G.R., Fowler, W.A. & Hoyle, F. (B^2FH, 1957) Synthesis of the Elements in Stars, *Reviews of Modern Physics* 29, 547–650; freely accessible via http://rmp.aps.org/abstract/RMP/v29/i4/p547_1

Chandrasekhar, S. (1939, new ed. 1958) *Introduction to the Study of Stellar Structure*, Dover, New York; freely accessible via http://archive.org/details/AnIntroductionToTheStudyOfStellarStructure

Larson, R.B. & Tinsley, B.M. (1978) Star formation rates in normal and peculiar galaxies, *Astrophysical Journal*, 219, 46–59; freely accessible via http://cdsads.u-strasbg.fr/abs/1978ApJ...219...46L

Maeder, A. (1992) Stellar yields as a function of initial metallicity and mass limit for black hole formation, *Astronomy & Astrophysics*, 264, 105–120; erratum (1993) *Astronomy & Astrophysics*, 268, 833; freely accessible via http://cdsads.u-strasbg.fr/abs/1992A%26A...264..105M

Renzini, A., & Voli, M. (1981) Advanced evolutionary stages of intermediate-mass stars. I — Evolution of surface compositions, *Astronomy & Astrophysics*, 94, 175–193; freely accessible via http://cdsads.u-strasbg.fr/abs/1981A%26A....94..175R

Tinsley, B.M. (1980) Evolution of stars and gas in galaxies, *Fundamentals of Cosmic Physics*, 5, 287–388; Note a small error discovered by Maeder (1992), p. 109.